How Old Is That Mountain?

How Old Is That Mountain?

A Visitor's Guide to the Geology of
Banff and Yoho National Parks

C.J. YORATH

ORCA BOOK PUBLISHERS

Canadian Cataloguing in Publication Data
Yorath, C.J., 1936
How old is that mountain?

Includes bibliographical references and index.
ISBN 1-55143-070-3

1. Geology–Alberta–Banff National Park–Guidebooks. 2. Banff National Park–Guidebooks. 3. Geology–British Columbia–Yoho National Park–Guidebooks. 4. Yoho National Park (B.C.)–Guidebooks. I. Title.
QE186.Y67 1996   557.123'32   C95-911173-5

Library of Congress Catalog Card Number 97-65294

Cover design by Christine Toller
Cover photograph by Ken McCormick
Interior photographs by C.J. Yorath unless otherwise indicated

Orca Book Publishers   Orca Book Publishers
PO Box 5626, Station B   PO Box 468
Victoria, BC V8R 6S4   Custer, WA 98240-0468
Canada   USA

99 98 97   5 4 3 2 1

To Linda

Table of Contents

PART TWO

FOREWORD

There they are, hunched over their desks and lab tables and computers at the Geological Survey of Canada: hundreds of people with Ph.Ds, discovering abstruse things about rocks.

These folks are spending taxpayers' dollars, and the stuff they do with the money is incomprehensible to most Canadians. Is this fair? Don't we have a right to know what's going on?

Of course we do, and the Geological Survey of Canada agrees. So they issue short annual reports writen in lay language and from time to time bring out non-technical publications on geological topics of general interest, such as the incredible fossils of the Burgess Shale.

Thus it was that GSC geologists Helen Belyea and David Baird produced a series of readable books about the geology of the national parks. I still have Baird's meaty little paperbacks on Banff, Jasper, Yoho and Kootenay, published between 1963 and 1968. They are fine books, full of wonderful photos of twisted rock layers, magnificent faults, overflowing glacial cirques, etc. And the texts explain everything. As a young geology student, I *loved* those books.

But 1968 was a long time ago. Progress in the science of geology has left Baird's and Belyea's books behind. To read them now is the job of the historian, not the person in quest of current knowledge.

Enter one Chris Yorath, formerly of the GSC's Pacific Division in Victoria. Among Dr. Yorath's many accomplishments, two bear strongly on this discussion: (1) he married a lady from Jasper who got him hooked on the Rockies, and (2) as he neared retirement he felt an increasing obligation to the non-geologist citizens who had paid his salary for 27 years.

This combination has spawned four books. The first was *Where Terranes Collide*, an entertaining account of the geological history and geological people of western Canada, published in 1990 and since made into an excellent film. The second, co-authored with myself, was *Of Rocks, Mountains and Jasper*, which came out in 1995 and replaced Baird's book on Jasper National Park. In the same year he produced *The Geology of Southern Vancouver Island*, co-authored with H. Nasmith. The fourth is the one you're holding, produced only two years later. This gives some indication of the speed and dedication with which Yorath has approached his post-GSC career.

We may conclude, then, that while the people of Canada have had to wait some years to get a proper update on the rocks of the Rockies, their bard has arrived. I hope you enjoy reading *How Old Is That Mountain?* as much as I did, and I hope that Chris gives us more of the same, complete with his whimsical asides and the odd line of quoted poetry.

Chris: how about a book on Kootenay National Park? And one on Waterton? And one on Glacier, and one on Pacific Rim, and one on Kluane, and ...

Ben Gadd
Jasper

PREFACE

How old is that mountain? How old are those ones over there? What are they made of? When were they made? And how? Have they always been there? Will they be there for ever?

This book is about the Rockies. More specifically, it's about the Rocky Mountains of Banff and Yoho national parks and adjacent areas. It's about thrust faults and glaciers, hoodoos and hot springs, carbonate and shale. And time. Vast amounts of time. To some extent it's also about people, about some of the geologists who spent much of their careers studying these mountains, reading the history that is entombed in their rocks.

HOW THE BOOK IS ORGANIZED

The book is organized in two parts. The first describes how geologists think these mountains came to be, what they are made of, and when they were made. Here I talk about the origin of the **Canadian Cordillera**, the system of mountain ranges and plateaus extending from the International Boundary with the United States to the Arctic

Ocean, and from the foothills of the Rockies to the edge of the continent off the west coasts of Vancouver Island and the Queen Charlotte Islands. In this section I discuss how the sedimentary rocks that form the Rockies developed, and how they were later deformed, uplifted, eroded and glacially sculpted into the mountains we see today. The second part is a tour through the Alberta foothills and Banff and Yoho parks. For the most part we will use the main highways, with occasional side trips to other points of interest such as Grassi Lakes, Lake Louise, Moraine Lake, Takakkaw Falls and Emerald Lake. For those of you who like to hike in the mountains I also describe the geology of the Lake O'Hara area.

On the inside of the front cover is the geological time scale, including the ages, in millions of years, of the boundaries between geological periods and eras. In parallel columns are the succession of major sedimentary rock layers, or **formations**, that occur in the two parks and adjacent foothills of western Alberta. By referring to the caption of this table you can get an idea of the main rock types of these formations. The position of each formation in the columns is in accordance with its age. The patterned gaps between formations indicate intervals of time for which rocks are not found in the two parks. In discussions of rocks and events older than 570 million years, I use the familiar and more general term "Precambrian" rather than the more correct term "Proterozoic," the youngest of three temporal subdivisions of Precambrian time.

Throughout the book are frequent references to the ages of rocks and times of events in terms of millions and hundreds of millions of years. Because such numbers are outside the normal temporal experience of most people, in the early pages wherever it seems helpful, by analogy I equate such times to the face of a twenty-four hour clock. For example, given that the beginning of the earth was 4.6 billion years ago, then this would correspond to time 0000 on the clock (midnight). The beginning of the Paleozoic Era, 570 million years ago, or 4.030 billion years after the formation of the earth, would correspond to a clock-time of about 2100 hours, or 9:00 PM. Using this analogy we can then say that when humans first appeared on the planet some 3 million years ago, the corresponding clock-time was 23:58:48 hours, or one minute and 12 seconds to midnight. I hope this analogy is useful.

Throughout the book geological jargon is kept to a minimum. However, because geology is a science of words it is difficult to avoid the use of technical terms. Besides, a lot of them are fun to use. When a technical term is first used, it appears in **bold** type and is used in explanatory context. Also, there is a glossary at the back.

One more introductory word. The intended audience for this book are those people interested in their natural surroundings. People who ask "How old is that mountain?" In this regard I urge readers of this book to purchase *Handbook of the Canadian Rockies,* by Ben Gadd of Jasper (*see Sources and Additional Reading*). Ben's *Handbook* is an outstanding compilation of the natural history of the Canadian Rockies, written in an easy, conversational style and well illustrated in full colour. If you want to know about geology, weather patterns, plants, animals, human history and recreation in the Rockies, this book has it all. Regarding the geology of the Rockies, the *Handbook* serves as an excellent support reference to what is said in this guide.

ACKNOWLEDGEMENTS

It is important to acknowledge that the information contained in this book was obtained from reports, scientific papers, as well as verbal communications with other geologists; nothing of what is said comes from my own work. Of particular importance are the scientific publications of Ray Price of Queen's University, Eric Mountjoy of McGill University and Jim Aitken, formerly of the **Geological Survey of Canada (GSC)**, and their many students and colleagues. In association with geologists of the Geological Survey of Canada, and of other universities, Ray, Eric and Jim undertook the detailed mapping and study of Banff, Yoho, Kootenay and Jasper national parks as part of the GSC's "Operation Bow-Athabasca" in 1965 and 1966. This was a helicopter-supported geological mapping project during which vast amounts of data were obtained on rock types, including their ages, distribution and structure. From these and subsequent studies over the past three decades our current understanding of the development of the Rockies has emerged. It is also important to acknowledge the work of the late R.J.W. Douglas of the GSC who, prior to Operation Bow-Athabasca, conducted extensive geological studies throughout the Rockies and established the structural framework upon which subsequent studies were based.

The many contributions made by geologists of the petroleum industry also added greatly to our knowledge of the Rockies. From the time of the discovery of oil and gas at Turner Valley south of Calgary, in 1914, petroleum industry geologists availed themselves of the rocks exposed in the Rockies so as to better understand the nature of their continuation beneath the surface of the prairies, where they were found to enclose many large pools of oil and gas. Of particular significance were the efforts of Shell Canada Resources Ltd., whose geologists car-

Ray Price of Queen's University (top) and Jim Aitken (bottom), retired from the Geological Survey of Canada, spent their careers studying Western Canada's mountain systems. It is these men, their colleagues and students to whom we owe our knowledge of the geology of the Candian Rockies.

ried out many detailed studies in the Rockies as part of their exploration programs and who shared their knowledge with universities and the Geological Survey of Canada. In like manner, many industrial geologists contributed information through the technical program of the Canadian Society of Petroleum Geologists which publishes the *Bulletin of Canadian Petroleum Geology* and sponsors many annual technical meetings, symposia and field excursions for its members.

I also wish to thank several people who directly contributed to this guide. My friend Don Cook provided the geological maps upon which much of this discussion is based. The Geological Survey of Canada kindly provided permission to use its illustrations, and Brian Sawyer and Richard Franklin of that organization assisted in their reproduction. My son Eric and Richard Franklin drafted those figures unique to this guide and provided photo-based drawings of geological phenomena. My publisher, Robert Tyrrell gave encouragement, support and a contract. I thank Bill Price, Ben Gadd, Ray Price, Trevor Lewis, Bruce Hart, Ken McCormick, Jim Aitken and Lee McKenzie McAnally for the use of their photographs. My wife Linda gave constant support and valuable editorial service. Nat Rutter and Peter Bobrowsky provided advice on those sections dealing with glaciation. Phillip Teece provided useful comments from the point of view of a general reader. Finally, I thank my good friend Jim Aitken who, true to his code, gave thoughtful and uninhibited criticism of the manuscript and provided many useful suggestions for its improvement.

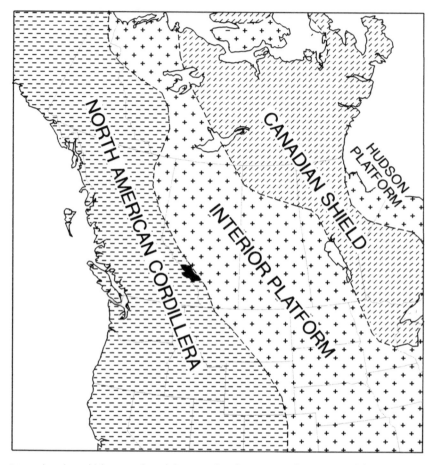

Principal geological/physiographic subdivisions of northwestern North America and the approximate locations of Banff and Yoho national parks. Natural Resources Canada, *Of Rocks, Mountains and Jasper* by Chris Yorath and Ben Gadd. Dundurn Press. Reproduced with the permission of the Minister of Supply and Services Canada, 1995.

PART ONE

THE GEOLOGICAL ARCHITECTURE AND ORIGIN OF THE CANADIAN ROCKIES

THE LAY OF THE LAND

Banff and Yoho national parks together embrace an area of 7,954 square kilometres. Banff National Park (6,641 sq. km), situated entirely within the province of Alberta, takes its name from Banffshire in Scotland, the birthplace of Lord Stephen, the first president and prime mover of the Canadian Pacific Railway. Its western boundary with Yoho (1,313 sq. km) and Kootenay (1,406 sq. km) national parks and Mount Assiniboine Provincial Park (385 sq. km), all in British Columbia, coincides with the Continental Divide separating Pacific from Atlan-

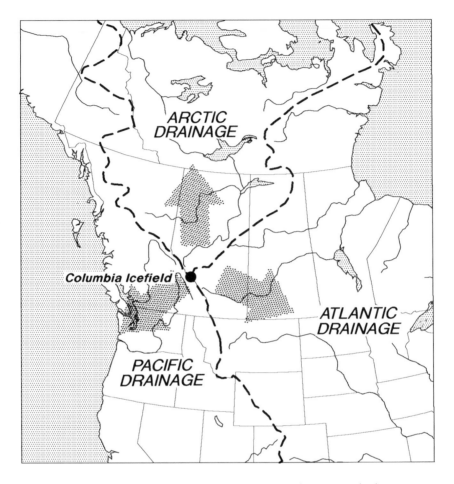

The two continental divides of North America, separating Pacific, Arctic and Atlantic drainage systems, intersect at the summit of the Snow Dome of the Columbia Icefield in Jasper National Park. This point is the hydrographic apex of North America and the focal point of the Pacific, Arctic and Atlantic drainage systems. *Natural Resources Canada, Of Rocks, Mountains and Jasper* by Chris Yorath and Ben Gadd. Dundurn Press. Reproduced with the permission of the Minister of Supply and Services Canada, 1995.

tic drainage. Its northern boundary with Jasper National Park (10,878 sq. km) lies within the Columbia Icefields, where it crosses the peak of Mount Athabasca just south of the Snow Dome (*see page 4 - 5*). The latter marks the hydrographic apex of North America where Pacific, Arctic and Atlantic drainage systems converge on a point where a mountaineer can contribute to three oceans in a single act.

East of Mount Athabasca the boundary traverses Sunwapta Pass, a watershed separating the Sunwapta River, a tributary of the Athabasca River (Arctic Drainage) from the North Saskatchewan River system draining from the Saskatchewan Glacier (Atlantic Drainage). The eastern boundary of Banff National Park follows a series of local drainage divides well inboard of the mountain front, from Nigel Peak in the north to near the headwaters of the Clearwater River; from there it swings eastward then southward in a broad irregular arc connecting several mountain front peaks as far as Ghost Lakes. From there it again turns inboard of the mountain front, joining several local drainage divides and ultimately meeting the Continental Divide on Mount Sir Douglas. Adjacent to its southeastern boundary is the recently established Peter Lougheed Provincial Park in Alberta.

One of Banff National Park's principal rivers is the North Saskatchewan, originating as meltwater from the Saskatchewan Glacier draining the Columbia Icefield. Before turning eastward to cross the **Front Ranges**, those ranges situated east of Castle Mountain, the Saskatchewan is joined by the Glacier and Howse rivers, originating from the Lyell, Mons and Freshfield icefields, and the Mistaya River flowing northward from Peyto Lake and Peyto Glacier. To the south the Bow River begins as meltwater from Bow Glacier, part of the Wapta Icefield, and the outflow of Bow Lake. From there the Bow flows southeasterly to cross the Front Ranges in a pair of dog-legs; along its southeasternmost reach, at the town of Banff, it is joined from the south by the Spray River, the Bow's principal tributary in the park.

In addition to those glaciers already mentioned which occur along and adjacent to the Continental Divide, Banff hosts several other large alpine glaciers, many of which supply meltwaters to nearby lakes. The colours of these lakes are commonly a spectacular emerald green, some darker than others due to higher concentrations of dissolved iron, and, more important, the presence of fine suspended particles, or **rock flour**. The particles of rock flour are a product of the erosive action of glaciers on bedrock and are transported to the lakes by meltwater streams draining from a glacier. Below a certain size, particles remain suspended in the water for many months and tend to backscatter those parts of the visible spectrum closest in wavelength to their average size, i.e. the blue- and green-end wavelengths, whereas other parts of the spectrum with longer wavelengths tend to be absorbed. In the spring, when snow-melt and run-off are greatest, the amount and size-range of sediment delivered to the lake is likewise greatest and the lake's colour becomes a muddy greyish green. In mid-summer, the total sediment input is reduced and most of that which reaches the lake is rock flour that remains suspended in the water. Thus, during the summer, the emerald-green wavelength of the spectrum is favoured in back-scattered light from the lake because the

GLACIERS AND ICEFIELDS

(A)	BOW GLACIER	(K)	LYELL ICEFIED
(B)	COLUMBIA ICEFIELD	(L)	MOLAR GLACIER
(C)	CROWFOOT GLACIER	(M)	MONS ICEFIELD
(D)	DALY GLACIER	(N)	PEYTO GLACIER
(E)	DRUMMOND GLACIER	(O)	PRESIDENT GLACIER
(F)	FRESHFIELD GLACIER	(P)	SASKATCHEWAN GLACIER
(G)	GLACIER DES POILUS	(Q)	VICTORIA GLACIER
(H)	GOODSIR GLACIER	(R)	WAPTA ICEFIELD
(I)	HANBURY GLACIER	(S)	WAPUTIK ICEFIELD
(J)	HECTOR GLACIER	(T)	WASHMAWAPTA ICEFIELD

M O U N T A I N S

▲ 1	ATHABASKA	▲ 20	MURCHISON
▲ 2	BOSWORTH	▲ 21	NIBLOCK
▲ 3	BORGEAU	▲ 22	NIGEL PEAK
▲ 4	BURGESS	▲ 23	NORQUAY
▲ 5	CASCADE	▲ 24	OUTRAM
▲ 6	CASTLE	▲ 25	RUNDLE
▲ 7	CASTLEGUARD	▲ 26	SARBACH
▲ 8	CATHEDRAL	▲ 27	SASKATCHEWAN
▲ 9	CHEPHREN	▲ 28	SIR DOUGLAS
▲ 10	CIRRUS	▲ 29	SNOW DOME
▲ 11	COLUMBIA	▲ 30	STEPHEN
▲ 12	CORY	▲ 31	SULPHUR
▲ 13	DENNIS	▲ 32	SURVEY PEAK
▲ 14	FIELD	▲ 33	TEMPLE
▲ 15	HEART	▲ 34	THREE SISTERS
▲ 16	INGLISMALDIE	▲ 35	WAPTA
▲ 17	ISHBEL	▲ 36	WILSON
▲ 18	LEFROY	▲ 37	YAMNUSKA
▲ 19	MISTAYA		

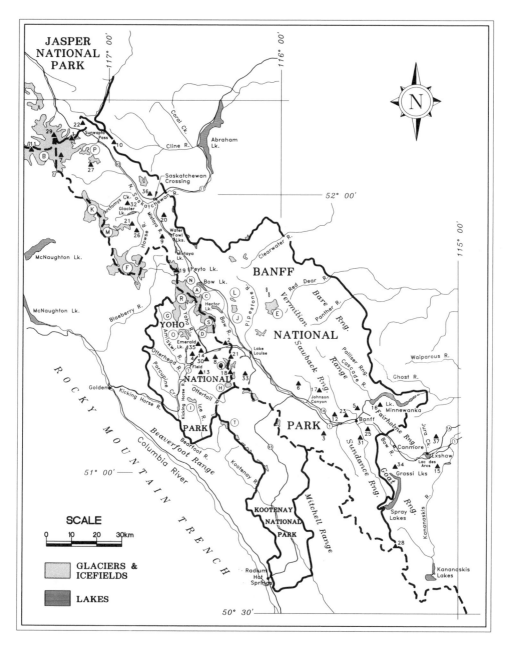

Map showing the principal mountains, rivers, glaciers and highways of Banff and Yoho national parks.

average size of the particles is closest to the wavelength of that part of the spectrum. In the fall, just before freeze-up, when particle concentrations are further reduced, the lake becomes clearer and bluer as progressively finer particles settle out of suspension to the bottom. Other controls on water colour are the depth of the lake and the presence of dissolved compounds of iron, copper and other elements. Interestingly, there is a remarkable uniformity of colour in Rocky Mountain lakes. This uniformity is thought to be due to particle dispersion by wind-driven turbulent flow in the uppermost levels of the water column. In this process the particles become uniformly mixed such that little or no variation in back-scattered colour is discernable. Some of the more spectacular examples include Lake Louise, Emerald Lake, Bow Lake, Hector Lake and Peyto Lake (*see pages 88, 105, 112, 113, 116*).

According to Philip and Helen Akrigg's book *British Columbia Place Names*, the name Yoho is taken from a Cree Indian expression of astonishment that roughly translates as "How Magnificent." Whatever its meaning, this comparatively small park does indeed display some of the most magnificent scenery in the Rockies. Its eastern boundary with Banff National Park coincides with the Continental Divide and its western boundary, beginning in the north at the Wapta Icefield, follows the drainage divide separating the Kicking Horse and Blaeberry drainage basins.

Yoho's principal river is the Kicking Horse which acts as a main trunk stream to tributaries such as the Yoho, Amiskwi, Ottertail, Otterhead and Beaverfoot rivers. The Kicking Horse River and the pass of the same name are thought to be named after an incident in August of 1859 when Dr. James Hector, a medical doctor and geologist with an expedition under the leadership of Sir John Palliser, was kicked in the chest by one of his packhorses. Kicking Horse Pass, near where the river heads, is on the Continental Divide between Mount Bosworth to the north and Mount Niblock to the south. Just west of the pass the Kicking Horse River begins as outflow of Wapta Lake from where it flows southwesterly to meet the Beaverfoot River then turns sharply northwesterly to pass into the Southern Rocky Mountain Trench at Golden where it empties into the north-flowing Columbia River.

As in Banff National Park, the main glaciers of Yoho are located along the Continental Divide. Those mainly within Yoho include Baker Glacier, Glacier des Poilus, Yoho Glacier, Diableret Glacier and Daly Glacier. Other glaciers occur on the President Range, Cathedral Mountain, Mount Vaux and Mount Goodsir.

The dominant style of landform in Banff and Yoho parks is one of northwesterly to southeasterly aligned parallel ranges and separating valleys. In Banff park the ranges most commonly have precipitous, east-facing slopes and less steep western slopes. The ranges are composed of erosion-resistant sedimentary rocks whereas the valleys have been carved into more easily eroded, softer sedimentary rocks. Of particular importance to the physiographic style of the Rockies, particularly in the Front Ranges, are the effects of weathering and erosion upon the rock strata exposed on the eastern slopes. At this latitude and in this climate, carbonates are much more resistant to the forces of weathering and erosion than are the softer, more easily erodible shales. It is for this reason that many of the mountains you see, such as Mount Rundle and Cascade Mountain (*see photo page 65*), consist of steep lower and upper carbonate cliffs separated by less steep slopes formed from limy shales. Geologists refer to these differences in the response to weathering and erosion as **differential erosion**; at this latitude, carbonate formations are said to be **resistant** as compared to shaly formations which are **recessive**. This juxtaposition of resistant and recessive formations lend a characteristic signature to the Rockies which will be further discussed in a later section.

Although the style of landforms in Yoho is similar to that in Banff, the mountains of Yoho lack the uniformity of shape displayed by the peaks of central and eastern Banff park, mainly because of the composition of their rocks and the manner by which those rocks reacted to the stresses that elevated them into mountains. Unlike the rock layers of the mountains of Banff, which most commonly are inclined to the west, the rock layering in Yoho lacks this uniform, westerly inclination.

A Few Basic Ideas

It is important to understand the distinction between mountain edifices and what they are made of. It's like a house, built of wood that was hewn from trees perhaps a hundred years old and yet the house itself may have been built only last month. Mountains are like that, built of rocks that were formed at one time and elevated into mountains at another time. When geologists look at mountains they must keep these two ideas separate in their minds. First they look at the rocks, their composition, the character of the layering, the nature of their contained fossils and so on. Then they look at the form of the mountain in which the rocks are exposed to view, its shape and internal structure. For example, the rocks of the Foothills and those of Banff and Yoho national parks were formed from sediments that accumulated over a period of some 710 million years; however, the forces

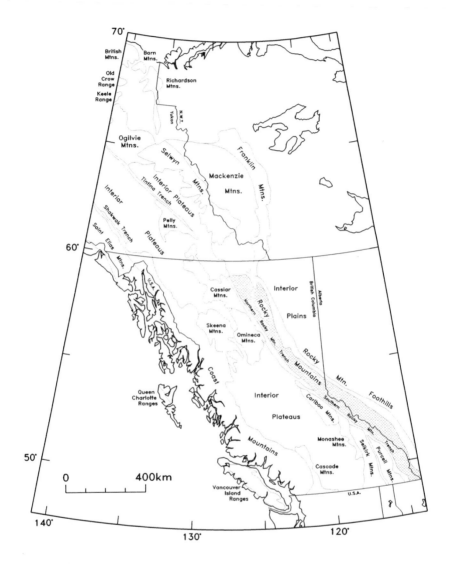

Physiographic divisions of the Canadian Cordillera. The several mountain systems occur as northwest/southeast-alligned groups of ranges, each having markedly different character depending upon their origin and rock composition. The Rockies (stippled), Mackenzies, Ogilvies and Richardson Mountains are composed of sedimentary rocks, whereas most of the other systems have been formed from sedimentary, igneous and metamorphic rocks. A broad region of plateaus separates the mountain systems of the eastern and western Cordillera. Natural Resources Canada, *Of Rocks, Mountains and Jasper* by Chris Yorath and Ben Gadd. Dundurn Press. Reproduced with the permision of the Minister of Supply and Services Canada, 1995.

that acted upon these rocks, elevating them into mountains, developed only between 120 and 60 million years ago. Once when I was a passenger on the Jasper Skytram, the young attendant was explaining to her charges that the mountain below us was such and such an age and the ones over yonder were of another age. It was the rocks in the two ranges that were of different ages, not the mountains. When I attempted to set her straight she replied, "Every time I get a geologist in this crate I get a different story." I have since remained silent on such occasions.

Another thing to realize is that the Canadian Rockies form only a part of the mountain systems of western Canada. Whereas some people believe that the name applies to all of our mountains, such is not the case. The Canadian Cordillera consists of many mountain systems, including the Rockies, Purcell Mountains, Selkirks, Cariboos, Skeena, Coast, Saint Elias and many other systems. As shown in the illustration on page 8, the Rockies include those ranges lying between the Foothills and the Rocky Mountain Trench. Forming the easternmost mountain system of the southern Canadian Cordillera, they extend from the International Boundary to the Liard River, close to the northern boundary of British Columbia, beyond which lies the broad arc of the Mackenzie Mountains in the Northwest Territories.

The rocks exposed in the Rockies are almost everywhere sedimentary rocks. Despite the assertions of those who speak impressively of the "granite wall of the Rockies" there are only a few known places in Banff and Yoho where such **igneous** rocks are exposed. We say "exposed" because granitic rocks do occur within the boundaries of the parks but at a depth of some nine kilometres below the surface. These rocks form part of the westward sloping surface of the **Canadian Shield** and as such form the **basement** upon and above which the layers of sedimentary rocks of the Rockies accumulated.

Another word for sedimentary layers is **strata**. Looking up at the cliffs of Mount Rundle and other mountains throughout the parks you will see that most are **stratified** or banded in different colours or hues. Some layers are thick, others are thin. These changes from one layer to another are due to variations in a wide variety of factors, including the composition of the sediments that make up the rocks which themselves are a products of the physical and/or chemical and/or biological environments which existed at the time the sediments accumulated.

The sedimentary rocks of the Rockies are of two main types: **carbonates** and **clastics**. Sedimentary carbonate rocks are composed of calcium and magnesium and/or iron in combination with carbon and oxygen. They most commonly are deposited from marine waters in tropical to subtropical settings. The most ubiquitous carbonates in the national parks are **limestone** and **dolomite**, the former composed of calcium carbonate ($CaCO_3$ — the mineral is called **calcite**) and the latter of calcium and magnesium carbonate ($Ca,MgCO_3$ — the mineral is called dolomite). All limestones in the Rockies formed through organic means, that is, through the ability of marine organisms such as algae, corals and brachiopods to extract calcium carbonate from sea water to build their colonies, shells and skeletons. Surprisingly, much limestone consists of tiny crystals of calcium carbonate formed within the cells of **cyanobacteria**, a kind of bacteria that has lived on earth for some 3.3 to 3.5 billion years. The carbonates of the Rockies are mainly composed of recrystallized lime mud in addition to the fragmented shells of marine organisms. In some cases whole shells of fossil organisms can be seen in the rocks; more often, however, the shells have been broken up by scavenging organisms and abraded and broken by marine currents to such an extent that only tiny fragments are seen. Good examples of fossiliferous limestone can be seen in outcrops alongside the many trails on Tunnel Mountain and in the decorative facades of some buildings in Calgary. The slabs on the building facades were quarried from the well-known Tyndall limestone of Manitoba, which contains easily recognizable large fossil shells of corals and other fauna embedded within finer particles of calcium carbonate.

Although most of the carbonate of the Rockies originally formed as limestone, some has been converted to dolomite. The process of **dolomitization** of limestone involves the partial substitution of calcium by the element magnesium in the atomic structure of calcite. In the process of dolomitization, whole fossil remains of organisms commonly are destroyed or replaced by other minerals. Furthermore, dolomitization leaves voids between the mineral grains and even large holes, or **vugs**, where fossil organisms had been entombed in the rock. Deep beneath the Alberta plains, fluids, including oil and gas, commonly are trapped within these porous, sponge-like rocks, thus forming many of Canada's oil and gas reservoirs.

Although the two forms of carbonate look very much alike, dolomite commonly has a distinctive sugary appearance. Even to geologists the two rock types are so similar that they need a means to tell them apart. To do this they carry in their shirt pockets a small plastic vial of dilute hydrochloric acid. When a drop of the acid is applied to limestone, the rock vigorously fizzes but when squirted on dolomite, the fizzing is very weak or absent. Geologists are thus easily recognized in crowds because their shirt pockets commonly are in shreds, having been half eaten away by leaking acid. Throughout the remainder of this book I will rarely distinguish between limestone and dolomite and simply call them carbonates.

Many outcrops of typical carbonates occur beside the Trans-Canada Highway east and west of Lac des Arcs east of the entrance to Banff National Park. In these outcrops you will notice that the light grey rocks have a blocky appearance imparted by **joints** or small fractures which often occur in sets of three: one set is parallel to the layering and the other two are vertical.

Clastic rocks are composed of particles, or **clasts**, of sand, silt, clay or rock fragments that have been eroded from other rocks and transported elsewhere by mechanical agents such as ocean waves and currents, rivers, winds, ice or landslide, etc. These kinds of rocks accumulate in nonmarine (terrestrial) or marine settings as **sandstone, siltstone,** and **shale**. Clasts larger than two millimetres in diameter form **conglomerate**. In Banff and Yoho national parks the minerals that most commonly comprise the grains of clastic rocks are **quartz** and **feldspar**. Quartz is a hard, insoluble (at surface temperatures and pressures) compound of silica and oxygen, capable of withstanding many cycles of erosion and deposition; whereas most other minerals such as feldspar are easily broken down during these processes and converted to **clays**, which commonly form the fine matrix of clastic rocks.

In the western part of the Rockies many of the oldest rocks have been **metamorphosed**: after having been subjected to great heat and pressure, many of the original minerals became altered to different compositions. For example, the clay minerals of shales commonly were changed to flakes of mica and green **chlorite**; the resulting rock thus tends to split, or cleave into thin slabs. These rocks we call **slate**. With increasing temperature and pressure the rocks become more intensely metamorphosed. The formation of **garnets** and other minerals indicative of increasing metamorphic grade, together with changes in texture, result in the formation of **schist** and **gneiss**.

Keeping in mind the distinction between the age of the rocks exposed in the mountain cliffs and the age of formation of the mountain structures themselves, it is also important to realize that the actual shape of individual mountains is due to the effects of erosion. As the land surface rises the forces of erosion immediately attack the rocks. Rivers carve deep valleys, while winds, rain, alternate freezing and thawing, biological activity and a host of other processes wear the rocks away. In the Rockies and many other mountain systems in the northern hemisphere, the greatest sculpting agent has been glacial ice. The enormous carving power of mountain glaciers over the past two million million years, together with aforementioned erosional processes have given us the mountain edifices we see today. In contemplating the valleys between mountains and mountain ranges, the broad valleys of the Bow, Saskatchewan, Athabasca and other rivers and their tributaries, it is not difficult to realize that since the Rockies began to be uplifted, about 120 million years ago, there has been more rock removed than currently remains. Where has all that rock gone? I'll get to that later.

You now have three basic concepts that are key to understanding the evolution of the Rockies. The first is that the rocks which are exposed in the two national parks and Foothills are old, deposited between about 750 million and 60 million years ago. Secondly, these rock strata were deformed and uplifted between 120 million and 60 million years ago; and the third concept is that the landforms, the peaks and valleys, were sculpted by erosion mainly during the past 60 million years, and by glacial ice during the past two million years.

ONCE UPON A TIME

The origin of the Canadian Rockies is intimately linked with that of other mountain systems of the Canadian Cordillera, the geological architecture and history of the Interior Plains, and to the origin and plate-tectonic regimes of the Atlantic and Pacific oceans.

Imagine that you are standing at the modern site of Banff Avenue about 750 million years ago. Muddy streams draining the vast granitic interior of the continent carry their loads of sand, silt and clay eroded from distant mountain ranges far to the east across a rolling and gently sloping surface. Not far to the west are other mountain ranges, lifeless, silent, unadorned. The air is cool. Glaciers are forming, becoming larger and coalescing into vast ice sheets. Earthquakes are frequent. To the west the land is being torn apart, rivers of lava issue from gigantic fractures and spread across the surface. Over several million years you notice that the mountain ranges to the west are receding, moving farther away. An ocean now separates and laps across the broken edges of the fracturing continent. Rivers carry huge volumes of sediment eroded from the eastern interior mountains to the sea. Over the next 180 million years, until the close of the Precambrian Eon, these sediments accumulate as a thick blanket across the jagged edge of the continent.

What you witnessed was the fragmentation of a giant

The supercontinent Rodinia as it might have looked prior to about 750 million years ago. Based on a drawing by P.F. Hoffman. Natural Resources Canada, *Of Rocks, Mountains and Jasper* by Chris Yorath and Ben Gadd. Reproduced with the permission of the Minister of Supply and Services Canada, 1995.

supercontinent and the beginnings of the geological history of Banff and Yoho national parks. By about 1.75 billion years ago, or about 3:00 PM on our time-clock, many of the world's primitive continents, or **Precambrian shields**, had amalgamated to form a giant continent we call **Laurentia**. This land mass included the Canadian Shield (North America), the Greenland Shield and possibly the Baltic Shield. Closely adjacent and perhaps welded to Laurentia along its western and northern margins were the East Antarctic, Australian and Siberian shields. By about 1150 million years ago Laurentia may have further collided with the South American and southern African shields to form a giant supercontinent which some geologists call **Rodinia**. Although the precise arrangement of the various continents within Rodinia is largely unknown, there is good evidence to suggest that the Precambrian shields of Australia and East Antarctica lay immediately adjacent to

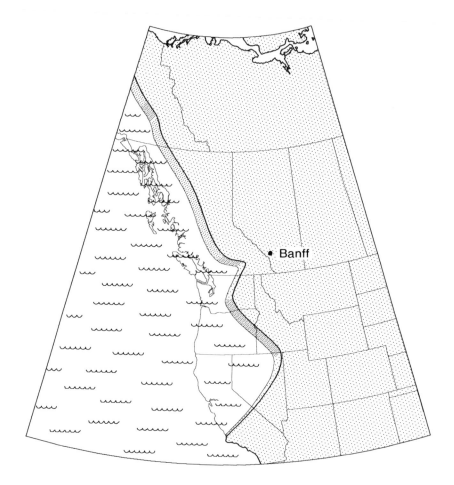

The edge of western North America about 750 million years ago, after Australia and East Antarctica had rifted away. Natural Resources Canada, *Of Rocks, Mountains and Jasper* by Chris Yorath and Ben Gadd. Dundurn Press. Reproduced with the permission of the Minister of Supply and Services Canada, 1995.

the western edge of North America. What you saw was the fragmentation of this supercontinent which had remained intact for about 400 million years before it broke up. The mountains to the west were those of Australia and East Antarctica.

The means by which Rodinia rifted apart is described by the theories of **seafloor spreading** and **plate tectonics**. In a manner similar to the opening of the Atlantic Ocean some 550 million years later, a **rift** developed, one side of which was more or less along a line joining the modern localities of Penticton, Kelowna, Prince George and points north and south. As sea floor was formed along the rift zone, North America drifted away from the rift in one direction and Australia and Antarctica in the opposite direction. In like manner, following the reassembly of the earth's continents during the late Paleozoic to form the supercontinent **Pangea**, North and South America rifted away from Africa and Europe, beginning about 200 million years ago, to form the modern Atlantic Ocean (*see page 29*).

Just as the opening of the Atlantic created a huge basin for the accumulation of sediments eroded from the separating continents, thus did the rifting of Rodinia create a basin for the accumulation of a thick blanket of sediments across the torn edge of western North America. The continental shelves forming the Grand Banks of Newfoundland and the Scotian Shelf are formed of sands and muds derived from erosion of the continent in just the same way as the Precambrian Miette **Group** (*of formations – see glossary*) of Banff and Yoho parks represents sediments deposited across the continental edge following the break-up of Rodinia.

The Miette Group comprises the oldest rocks you see in the two national parks and is well exposed in roadside outcrops along the Trans-Canada Highway from Castle Junction to well past Lake Louise. Here the outcrops are assigned to the Corral Creek and Hector formations of the Miette Group and consist mostly of dark grey, green and purple slate and sandstone; elsewhere, conglomerate and limestone form part of this succession of strata. The Miette Group and other closely related sedimentary rocks form a succession of strata ten kilometres thick that was deposited across the torn edge of the continent during the latter part of the Precambrian Eon, between about 750 and 650 million years ago.

Before continuing I should note that the Miette Group is not the oldest succession of sedimentary strata in the Canadian Rockies. In Waterton Lakes National Park in southwestern Alberta and adjacent Glacier National Park in the United States, the mountains are composed dominantly of clastic and lesser carbonate strata of the Purcell

The Corral Creek Formation of the Precambrian Miette Group is exposed as steeply inclined slates at the intersection of the Trans-Canada Highway and Icefields Parkway. (Photo by W.R. Price.)

Supergroup (yes we geologists have supergroups as well). The Purcell Supergroup consists of several formations which appear to have been deposited in a small oceanic basin that became trapped between the continents of Laurentia (i.e., North America) and East Antarctica and Australia when they first assembled as the beginnings of Laurentia/Rodinia about 1750 million years ago. Between then and the time of the break-up of Rodinia, about a billion years later, up to twenty kilometres of strata had accumulated in the trapped basin, which appears to have had similarities to the modern Caspian Sea. The rocks in Waterton Lakes National Park are between 1700 and 1200 million years old and are not represented by surface exposures in either of Banff or Yoho national parks.

Back to our story. Because of its place of accumulation along the edge of the continent the Miette Group exhibits many of the characteristics of modern sediments currently being deposited on northern continental shelves, continental slopes and adjacent deep ocean floors. Many of these and related types of deposits are classed as **turbidites**, which are deposited from dense, sediment-laden submarine currents, commonly formed from earthquake-induced landslides along the steep outer continental slopes.

At the time of the actual break-up of Rodinia much of the supercontinent was covered with one or more thick sheets of glacial ice. As with other **glaciations** such as those during the so-called "Ice Age" (which we may still be experiencing), global sea levels were lowered as more and more water was withdrawn from the oceans to form the ice sheets. During late Precambrian time the Miette picture may have been analogous to that of the modern Ross Ice Shelf which extends outwards over the surface of the sea off the coast of Antarctica. As the glacier moves outward from continental Antarctica the debris eroded from the continent becomes entrained in and on the ice. Upon reaching the sea the buoyant ice floats and moves progressively farther seaward until it begins to break up and melt, at which time the debris frozen in the ice falls to the muddy sea floor to become incorporated in the bottom sediments as coarse pebbles, cobbles and boulders. The resulting mixture of coarse debris enclosed in fine mud is called a **diamictite**, of which the Miette Group has several examples, particularly near its base. These same kinds of deposits have been found in equivalent rocks in Australia and China, which suggests that just as the supercontinent was beginning to break up it was, at the same time, being glaciated. Following break-up the fragments of the supercontinent drifted apart, during which time the Corral Creek and Hector formations accumulated. The lowermost part of the Hector Formation, where it overlies the Corral Creek Formation, consists of limestone-bearing strata which Gerry Ross of the GSC thinks reflect a rising of sea level following the melting of the ice sheets. These same deposits have been noted elsewhere around the world in rocks of the same age, lending further support to the notion of widespread continental glaciation during late Precambrian time.

LEGACY OF A KICKING HORSE

Returning to our stroll along the future site of Banff Avenue, the time is now some 570 million years ago, about 9:00 PM on our clock. Australia and Antarctica have drifted far away. The continental shelf, formed by strata of the Miette Group, is covered by a shallow sea encroaching landward from the widening deep ocean to the west. To the east, toward the interior of the continent, the remaining Precambrian mountains by now have been eroded to little more than hills drained by river systems which continue to carry their loads of sediment westward to the shallow sea where they accumulate as sandstones of the Lower Cambrian Gog Group.

Whereas the Precambrian seas were virtually devoid of life, except for algae and soft-bodied animals without shells such as worms and primitive jelly fish, the seas of the Cambrian Period at the beginning of the Paleozoic Era teemed with life. **Trilobites**, those oblong, segmented creatures with names like *Nevadella, Olenellus and Fremontia,* skittered about on the muddy bottom. In some areas protected from the invading clastic sediments and where the waters were warm and clear, sponge-related creatures with ice-cream cone-like shells called **archaeocyathids** built carbonate reefs upon limestone platforms formed from the carbonate secretions of marine algae and accumulated broken shell material. One locality is particularly noteworthy. Many years ago Jim Aitken gave the name "**Kicking Horse Rim**" to an ancient physiographic feature across which rocks of Middle and Late Cambrian age change in character, or **facies**, from types that were deposited in shallow water to those which accumulated in deeper water. In the vicinity of Field, B.C. this feature may have been a steep submarine cliff of carbonate, off the top of which fine muds and myriads of bottom-dwelling creatures may have been swept by storm-induced currents, thence to settle in deep, oxygen-depleted waters at the base of the cliff. The low oxygen content of the water would have allowed the soft tissues of these creatures to become buried and preserved in the mud before decay could set in and destroy them as it does in most other cases. As a consequence the Burgess Shale, a part of the Cambrian Stephen Formation, on the west slope of Mount Field in Yoho National Park, is one of the most important fossil localities in the world. This locality was first investigated in 1910 by Charles D. Walcott, Director of the Smithsonian Institution in Washington, D.C. and the most eminent American scientist of his time. It has since been declared a World Heritage Site by the United Nations. The importance of the Burgess Shale is that it not only contains outstanding preservations of the soft tissues of 550 million-year old organisms, but also that it contains the record of the "Cambrian explosion" of an incredible diversity of multicellular life. In addition to trilobites the Burgess Shale contains several other kinds of arthropods, as well as various kinds of worms, echinoderms, sponges and other creatures. In fact, the Burgess Shale contains representatives of all of our major animal groups that are living in our modern oceans as well as several other groups which have long been extinct. An excellent illustrated publication by the Geological Survey of Canada entitled *Fossils of the Burgess Shale: A National Treasure in Yoho National Park, British Columbia* is available through the GSC's offices in Calgary, Vancouver and Ottawa (*see Sources and Additional Reading*). For further discussion of the fossils of the Burgess Shale see page 101. I also heartily reccommend the book *Wonderful Life* by Stephen Jay Gould of Harvard University (*see Sources and Additional Reading*).

Trilobites such as *Ogygopsis klotzi,* collected from Mount Stephen (top), together with many other arthropods such as *Aysheaia* (bottom), a fossil of the Burgess Shale, were bottom-dwelling, scavenging, marine organisms during the Cambrian time.

Following deposition of clastic strata of the Gog Group during the Early Cambrian the remainder of Cambrian time saw the further encroachment of the sea eastward onto the continent. In the shallow waters of this broad shelf alternating formations of carbonate and shale accumulated, whereas in deeper waters to the west limy shales were deposited. This change in facies, reflecting changes in depositional settings from east to west, is characteristic of all the Paleozoic rocks along the entire length of the Rockies and Mackenzie Mountains of the Northwest Territories. Apart from variations in local submarine topography, such as on either side of the Kicking Horse Rim, the broad picture during Cambrian and later time was a gently westward sloping shelf, covered by a gradually westward deepening sea; carbonates accumulated in the east and shales to the west. This shelf and the sedimentary strata that accumulated upon it throughout Paleozoic and early Mesozoic time are referred to as the **miogeocline**.

Rocks of Cambrian age occur widely throughout Banff and Yoho national parks. Some of the more impressive mountains containing Cambrian strata are Mount Temple and the mountains surrounding Moraine Lake (*see photo page 65*) and virtually all of the mountains seen from the Trans-Canada Highway between Lake Louise and Golden and along the southern part of the Icefields Parkway as far as Saskatchewan Crossing (*see page 116*). It is formations of Cambrian age that best display the change in facies from east to west across the Rockies. The eastern facies, characterized by shallow-water carbonate, includes the Cathedral, Stephen, Eldon, Pika, Arctomys, Waterfowl, Sullivan, Lyell, Bison Creek and Mistaya formations. To the west, beyond Field, B.C., Cambrian strata are dominantly dark limy shale and muddy limestone contained within the Chancellor Formation; however, a thick prominent carbonate, called the Ottertail Formation, also occurs within the western facies. Excellent examples of eastern carbonate strata are seen in the Eldon Formation, forming the impressive cliffs of Mount Yamnuska east of the town of Exshaw (*see photo page 71*), and in Castle Mountain where the Cathedral, Stephen and Eldon formations make up the battlements of this massif (*see cover photo*). Near the town of Field you can see the change in facies between the upper part of Mount Stephen and the lower part of its neighbour to the west, Mount Dennis. The upper cliffs of the former expose carbonates of the Eldon Formation, whereas the comparatively subdued topography of the lower slopes of Mount Dennis reflects the more easily eroded limy shales and muddy limestone of the Chancellor Formation (*see page 101*). Another example of this change can be seen at Emerald Lake where the light-coloured carbonates of the Eldon Formation on

Wapta Mountain change to dark limy shale of the Chancellor Formation on Mount Burgess (*see page 106*). Each of these examples occur along the the Kicking Horse Rim.

An interesting formation of the Cambrian System is the Arctomys Formation. Although not well exposed at localities easily accessible to the tourist, the Arctomys can be seen as a comparatively thin formation of steeply inclined strata composed of brownish-orange siltstone, shale, carbonates and **evaporites** on the south shoulder of the Sawback Range as viewed northward from the Trans-Canada Highway west of Banff, or better still, from atop the gondola on Sulphur Mountain (*see page 83*). There the recessive Arctomys is bracketed by prominent ribs of cliff-forming carbonates forming the younger Waterfowl Formation to the west and the older Pika Formation to the east. It is believed the Arctomys represents a period of alternating environments of extremely shallow water followed by exposure and drying of the miogeocline, some 530 million years ago. As the salty marine waters evaporated, thin layers of limestone, dolomite, anhydrite and gypsum containing large salt crystals were precipitated as products of the evaporation. The siltstones also are interesting. Rein DeWit, a well known petroleum geologist, now retired, studied many exposures of the Arctomys and concluded from examination of the grains of the siltstones that they appear to have been deposited by wind-driven dust storms blowing across the exposed shelf surface. This kind of material is called **loess**. Thus, for a short time some 530 million years ago, the ancient Cambrian shelf was a desert until the sea readvanced to deposit the carbonates of the Waterfowl Formation.

The maximum total thickness of shallow-water strata deposited during the Cambrian Period was almost 5000 metres, which leads to the question of how did such a great thickness accumulate in water depths of 150 metres or less? As sediments accumulate, the **crust**, or outermost layer of the earth, subsides under the weight of the sediment load thereby allowing for continued accumulation. At and near the edges of continents, not only do sediments build up to great vertical thicknesses but they also accumulate progressively farther seaward toward the deep open ocean floor. Another factor contributing to sediment accumulation is thermal cooling and sinking of the crust. From the time of initial rifting of Rodinia, the crust along the ancient edge of the drifting continent began to cool, thereby becoming less buoyant and consequently sinking, or subsiding, progressively deeper below sea level, thus permitting more sediment to accumulate.

By Late Cambrian and Early Ordovician time shallow marine waters had spread over much of the continent but withdrew for a short interval only to expand again throughout the remainder of Ordovician and Early Silurian time. During these periods of marine advance onto the continent, life continued to increase in both numbers and diversity. Brachiopods, corals and bryozoans thrived on the sea floor while above them colonial organisms called **graptolites**, now long extinct, floated about on the surface.

Like Cambrian strata, Ordovician rocks also show the east-to-west change in facies across the miogeocline. Eastern exposures, such as those seen on the imposing edifice of Mount Wilson at Saskatchewan Crossing along the Icefields Parkway, consist of limestones, shales, dolomites and sandstone of the Survey Peak, Outram, Skoki, Owen Creek, Mount Wilson and Beaverfoot formations. To the west, in the region of Kicking Horse Pass west of Lake Louise, Ordovician rocks are assigned to the upper part of the McKay Group and Glenogle Formation. The McKay is easily recognized as the light grey, greenish grey and yellowish grey contorted shaly rocks along the narrow and twisting part of the Trans-Canada Highway through Kicking Horse Canyon. In the same area the Glenogle Formation is a black shale containing abundant graptolites. To the determined hunter armed with a hand lens, these can be seen as shiny, leaf-like black films on bedding surfaces of shale **talus** which has spalled from outcrops at the west end of the first bridge on the Trans-Canada Highway over the Kicking Horse River west of Field (*see page 107*).

TIME LOST

During the latter part of the Silurian Period the seas withdrew from the southern miogeocline in the vicinity of the two parks. This withdrawal was in response to widespread uplift of the miogeocline whereby the western part of the continent emerged above sea level and was subjected to erosion. In some places the rocks deposited during the Ordovician and latter part of the Cambrian periods were eroded away. It wasn't until the latter part of the Devonian Period, some 45 million years later, that the sea readvanced across the southern miogeocline. Intervals of time between withdrawals and readvances of the sea, when no sediments are deposited and when the forces of erosion attack the newly exposed land, are marked by **unconformities**. These are boundaries separating strata of markedly differing ages such that between the youngest rocks beneath the unconformity and the oldest strata above there is no geologic record for the missing interval. These episodes of marine advance and withdrawal can be caused by one or more of the following processes: uplift or subsidence of the land surface; by the rise and fall of sea level due to plate tectonic processes, including the growth and destruction of mid-ocean ridges; or by the waxing and waning of continental ice sheets that remove large volumes of water from the oceans. In the first of these processes large thermal convection cells in the **mantle** deep beneath the crust cause the land surface to rise or fall such that from region to region the land surface may be rising in one area and subsiding in another. Such was the case during Silurian and Early Devonian time when the sea withdrew from the south but remained covering the northern miogeocline resulting in widespread deposition of carbonates throughout northern British Columbia and the Mackenzie region.

A good example of an unconformity is visible along the mountain front just east of Exshaw (*see page 73*). Looking north as you approach the mountain front from the east, you can see the prominent cliff of Mount Yamnuska. As shown in the photograph on page 73 this cliff, together with those on adjacent mountains are composed of carbonates of the Middle Cambrian Eldon Formation. Looking to the mountains to the south (toward you) of Mount Yamnuska you can see a thin, yellowish zone above the Pika Formation which overlies the Eldon. This thin zone separates the Cambrian formations below, from Devonian formations of the Fairholme Group above, the latter extending up to the skyline. Missing between them are Ordovician, Silurian and Lower and Middle Devonian strata. The yellow zone represents a period of final weathering prior to the readvance of the sea during Late Devonian time and thus marks an unconformity, expressing the absence of any record of the earth's history in this region for some 160 million years (about 50 minutes on our time clock)!

THE GOLDEN EGG

The readvance of the shallow sea across the southern miogeocline during the latter part of the Devonian Period, about 375 million years ago (10 PM), heralded the onset of widespread deposition of carbonate strata and the development of chains of organic reefs that were ultimately to play an important part in the economic development of western Canada. Alberta towns such as Leduc and Swan Hills owe their existence to the fact that, buried some 1,500 to 2,500 metres below the surface, these carbonate rocks contain some of Canada's most important oil and gas reserves. In the eastern part of Banff National Park and beyond to the mountain front these carbonates, together with Lower Carboniferous strata, are magnificently exposed in the pale blue-grey cliffs of the Front Ranges of the Rockies.

This cairn, located in front of the Geological Survey of Canada's offices in Calgary, is a piece of a Devonian limestone reef collected from rocks in the Front Ranges of the Rockies. Deep below the surface of Alberta, the interconnected pores of these reefs are filled with oil and gas.

The Devonian strata of the Rockies are contained within several component formations of the Fairholme Group as well as the Alexo, Sassenach and Palliser formations. In the Front Ranges of Banff National Park the Fairholme Group includes, in upward order, the Flume, Cairn and Southesk formations, all of which are carbonates. Locally, beneath the Flume Formation, are ancient stream channels that were cut into the underlying Cambrian shelf strata and filled with clastic river deposits of the Yahatinda Formation. The Yahatinda sediments represent the first deposits which accumulated upon the miogeocline since the withdrawal of the Silurian shallow seas, some 50 million years earlier. With the return of marine conditions to the miogeocline, vast tracts of carbonates accumulated in the clear, warm shallow seas where myriads of marine organisms extracted calcium carbonate from the sea water to build their shells and construct broad carbonate banks. In some areas where circulation patterns and water clarity permitted, algae, corals, brachiopods and creatures called **stromatoporoids**, extinct organisms thought by many paleontologists to be related to sponges, built chains of organic reefs on top of the carbonate banks. One such reef is magnificently exposed in a lovely little park at Grassi Lakes above the town of Canmore (*see page 76*).

The Devonian carbonate reefs exposed in the Rockies and buried beneath the plains of Alberta are surrounded and overlain by black, limy shales rich in organic matter. In the mountain parks these, together with muddy limestones containing abundant clastic material, are included in the Perdrix and Mount Hawk formations, the so-called "off-reef" facies of the Fairholme Group. "Off-reef" is geologese for those areas beyond the reefs where the shallow sea was turbid and muddy, resulting in the accumulation of clastics mixed with carbonate. In Banff National Park neither the recessive Mount Hawk Formation nor the Perdrix are exposed in readily accessible places and thus are not further discussed in this book. Likewise, the Alexo and Sassenach formations, together consisting of a succession of mixed carbonate and clastics that occur above the Fairholme Group, are not well exposed because of their recessive character. However, good exposures of the Alexo Formation can be seen in the roadside outcrops along the Trans-Canada Highway on the south side of Lac des Arcs.

During the time of Alexo and Sassenach formations accumulation a catastrophic climatic event of global proportions is thought to have virtually wiped out all reef-building organisms such as those which had constructed the reefs of the Fairholme Group. In 1982, Digby McLaren (GSC, retired) proposed that the cause may have been the impact of a large meteorite. Such events are believed to have caused enormous amounts of dust to be blown into the atmosphere, ultimately leading to a substantial reduction in the amount of the sun's radiation reaching the earth's surface — the so-called "nuclear winter" effect feared during the cold war years. If this indeed happened, the result would certainly have been a catastrophic reduction in animal life on the planet. It is precisely this cause which is believed to have wiped out the earth's dinosaurs at the end of the Cretaceous Period, some 300 million years later. Perhaps the sandy parts of the Sassenach Formation may, in part, represent "fallout" from a globe-encircling dust storm, possibly caused by a meteorite impact.

In the Front Ranges of Banff and Jasper national parks the most prominent formation of Devonian age is the Palliser Formation, named for Sir John Palliser, a British explorer of western Canada during the nineteenth century. Like the Upper Cambrian Eldon Formation, from which it is distinguished with difficulty, the Palliser is a prominent cliff-forming mass of grey carbonate which makes up the lower, precipitous cliffs of such mountains as Rundle, Cascade and Inglismaldie near Banff and the Weeping Wall of Cirrus Mountain near the northern boundary of Banff National Park (*see pages 26 and 119*). The Palliser represents a return of clear, warm subtropical marine environments to the miogeocline following the earlier climatic catastrophe.

The Weeping Wall forms the lowermost cliffs of Cirrus Mountain where several small streams cascade over gently inclined carbonate strata of the Devonian Palliser Formation.

The next youngest formation and the uppermost of Devonian age is an interesting succession of black, bituminous shale, as well as siltstone and minor limestone called the Exshaw Formation. Because it is only about fifty metres thick and made of soft and easily weathered materials, together with the fact that it is overlain by a thick formation of calcareous shale of similar colour called the Banff Formation, it is not readily identifiable on distant mountain faces in Banff National Park. The lowermost ten metres of the formation have a very high organic content and are thought to have accumulated in moderately deep, poorly oxygenated water where the organic material was preserved. The intriguing aspect of the lower Exshaw is that it occurs throughout an area of over seven million square kilometres in central North America where it is known by a variety of names, including the Chattanooga Shale. How such uniform, deep water environments could have suddenly developed following widespread shallow water conditions, and over such a huge area of the continent at the same time, has not been convincingly explained. Another interesting aspect of the Exshaw is that its organic content is thought to have been the source material for oil and gas pools trapped in Carboniferous carbonate reservoirs in the subsurface of western Alberta; the Turner Valley oil and gas field southwest of Calgary is a good example. If you wish to see an exposure of the Exshaw Formation, a walk of about three kilometres northward along Jura Creek, about one-and-a-half kilometres east of Exshaw, should satisfy you.

In earliest Carboniferous time, following deposition of the Exshaw Formation, the continental shelf of the southern miogeocline was the site of the accumulation of limy muds and muddy carbonates of the Banff Formation. The clastic materials in the Banff are thought to have been derived through erosion of far distant uplands from where the eroded particles delivered to the shelf by streams and winds where they settled out of the sea to mix with carbonate. The Banff Formation is easily identified on many mountains in the Front Ranges where it forms the recessive middle slopes above the Palliser Formation of Mount Rundle, Cascade Mountain and many others. Above the Banff and forming the impressive uppermost cliffs of these mountains is the Livingstone Formation, the lowermost subdivision of the Rundle Group. The Livingstone marks a return of crystal clear tropical to subtropical seas to the miogeocline. The formation is composed mostly of the broken shells of echinoderms and bryozoa which probably accumulated in environmental settings much the same as those observed today in the Qatar Peninsula area of the Persian Gulf. The Carboniferous limestones of the Livingstone represent broad carbonate banks which periodically were subjected to storms, resulting in the destruction and abrasion of shell material and its subsequent distribution over thousands of square kilometres by the currents of this shallow sea. The end result was the accumulation of a succession of carbonate strata to which the Rockies owe much of their spectacular scenery.

Above the Livingstone Formation are carbonate and clastic strata of the Mout Head and Etherington formations, the upper two subdivisions of the Rundle Group in the Banff area. These strata acumulated during several episodes of marine advance and retreat across the southern miogeocline during a time when the world's continents were again reassembling into another giant supercontinent.

In Banff National Park the three-fold succession of Devonian and Lower Carboniferous formations lends a unique signature to the physiography or shape of the mountains which is widely recognizable throughout the Front Ranges. The cliff-forming Palliser and Rundle Group, separated by the recessive, gentler slope-forming Banff Formation, occurs in each of the five northwesterly-aligned parallel ranges from the mountain front to the Sawback/Sundance ranges west of Banff. The reason for this repetition in formations and physiography in range after range has to do with thrust faults. More on that later.

PANGEA, OH PANGEA

It was during the Carboniferous Period, some 300 million years ago, when the world's continents again reassembled into a giant supercontinent we call Pangea. At about this time carbonate and clastic strata of the Spray Lakes Group were accumulating upon the miogeocline, following which the sea withdrew until the Early Permian when a readvance of the sea allowed for the deposition of shales and carbonates of the Ishbel Group. An interesting aspect of the Ranger Canyon Formation, the uppermost subdivision of the Ishbel Group, is its content of **phosphorite**, composed of the mineral **apatite**, or calcium phosphate. Deep ocean-bottom waters are essentially saturated in calcium phosphate which, when subjected to slight changes in physical and chemical conditions, will precipitate out of solution, particularly when little other sedimentary material is accumulating. Such conditions commonly occur at the margins of continents where upwelling of ocean-bottom waters results in the transfer of phosphorous-rich waters to the shallow continental shelves. During the Permian and Triassic periods, southward-directed currents along the western margin of North American Pangea possibly caused such upwellings to be frequent enough to allow for extensive phosphate deposits to accumulate. The Phosphoria Formation of Montana, Idaho and Washington contains extensive economic phosphate deposits and is the equivalent of the Ranger Canyon Formation of the Ishbel Group. A good exposure of the Ishbel Group occurs on the west slope of Mount Norquay, where together with Carboniferous and Devonian rocks, it forms a prominent, westerly inclined succession of carbonate and clastic strata on the north side of the Trans-Canada Highway.

The end of the Paleozoic Era was marked by another widespread and devastating mass extinction when perhaps as much as 90 percent of marine invertebrate species and between 75 percent and 80 percent of all amphibian and reptile groups as well as many plant and insect species suddenly disappeared. Although the cause or causes of this are uncertain, the assembly of Pangea leading to the withdrawal of the shallow seas from the supercontinent may have been an important factor. With a dramatic reduction in the shallow shelf areas surrounding the continental interiors, the ability of marine organisms to maintain their diversity and abundances would have been greatly curtailed. At the same time the development of widespread arid climates led to extensive deposition of salt through evaporation, with a resulting decrease in ocean salinity. It has been suggested that removal of salts from the oceans would have lowered salinities to lethal levels for

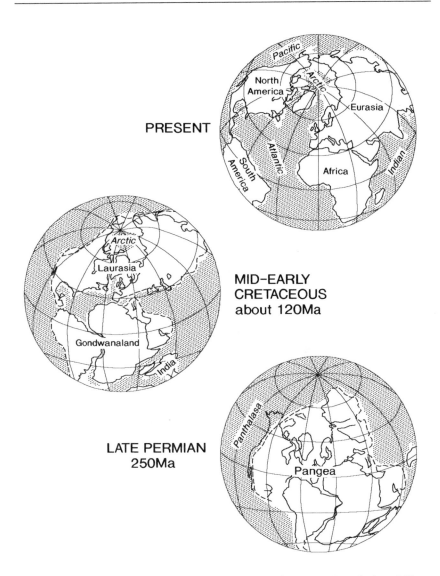

PRESENT

MID–EARLY
CRETACEOUS
about 120Ma

LATE PERMIAN
250Ma

Three stages in the opening of the Atlantic Ocean. During the Permian Period, about 250 million years ago, the globe's continents were gathered together into one giant supercontinent called Pangea. Beginning about 180 million years ago Pangea disassembled into two huge continental masses separated by a widening Atlantic Ocean. To the north was Laurasia, consisting of North America and Eurasia, and to the south was Gondwanaland, made up of South America, Africa, Antarctica, Australia and India. The present configuration of the continents is shown at the top; India was added to Eurasia about 40 million years ago. Based upon drawings by E. Irving. From *Where Terranes Collide* by C.J. Yorath. Orca Book Publishers. Ma = millions of years ago.

many marine organisms. Another recently suggested cause may have been a period of continuous volcanic eruptions which lasted for perhaps as long as a million years. Massive amounts of sulphur dioxide, carbon dioxide and other gasses injected into the atmosphere by these eruptions could have blocked sunlight and rendered the air so foul so as to eventually make the planet virtually uninhabitable.

The supercontinent Pangea persisted for about 120 million years, during which time marine shale, siltstone and limestone of the Triassic Spray River Group (not to be confused with the Spray Lakes Group) accumulated following the brief marine withdrawal from the miogeocline near the close of the Paleozoic Era. Strata of the Sulphur Mountain Formation of the Spray River Group are seen in outcrops along the road leading up Mount Norquay and form the cataract of Bow Falls (*see page 82*). At this time Pangea was surrounded by a huge ocean which geologists call **Panthalassa**.

The break-up of Pangea and the initial stages of formation of the modern Atlantic Ocean began during the Early Jurassic Period, about 175 million years ago. Pangea first separated into two large chunks, a northern mass called **Laurasia** which included North America and most of Europe and Asia, and a southern megacontinent called **Gondwanaland**, including South America, Africa, Antarctica and India. As the two megacontinents separated the central Atlantic formed first and became connected to a narrow seaway known as **Tethys**, which extended eastward across southern Europe. With the subsequent opening of the South Atlantic about 130 million years ago, the world's continents began to move toward the arrangement they have today. The further break-up of Gondwanaland ultimately led to the northward migrations of Africa and India, the former colliding with southern Europe to create the Alps throughout the Mesozoic and Early Cenozoic eras, and the latter colliding with the underside of Asia about 40 million years ago to form the Himalayas.

The processes of continental break-up, formerly embodied under a general hypothesis called **continental drift**, today are embraced by the two related theories of seafloor spreading and plate tectonics. The creation of seafloor spreading ridges, such as the Mid-Atlantic Ridge, in addition to being the locus of generation of new oceanic crust, also can cause continents to split apart when such ridges are generated in the mantle beneath them. For example the Red Sea contains a spreading ridge where new sea floor has been forming between the Arabian Peninsula and Africa for the past 25 million years. As spreading continued a rift has propagated through the Gulf of Aqaba and along the valley of the Dead Sea, the surface of which is almost 400 metres below sea level.

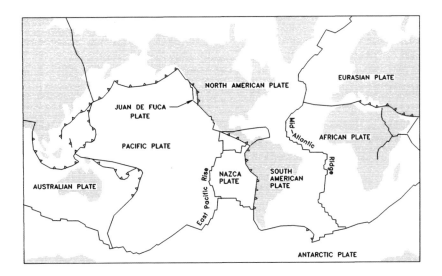

The earth's outer shell, including the crust and upper part of the underlying mantle, is divided into several independantly moving plates that are separated from one another by seafloor-spreading ridges, faults and deep sea trenches (top). The seafloor-spreading ridges, such as the Mid-Atlantic Ridge, the East Pacific Rise and the Juan de Fuca Ridge, are zones along which new seafloor is created and added to the edges of the plates. As new material forms, the plates spread away from the ridges toward the deep sea trenches where they are consumed, or subducted, beneath adjacent plates. The current plate-tectonic situation off the west coast of Canada is shown below. Molten magma from the mantle is injected into the Juan de Fuca Ridge, creating the Juan de Fuca Plate to the east and the Pacific Plate to the west. As the Juan de Fuca Plate spreads eastward, it is subducted beneath the westward moving North American Plate. At a depth of between 150 and 200 km the oceanic crust melts, whereupon the molten material rises upward to appear at the surface of the overriding plate as the Cascade Volcanic Arc. Based upon drawings by R.P. Riddihough. From *Where Terranes Collide* by C.J. Yorath. Orca Book Publishers.

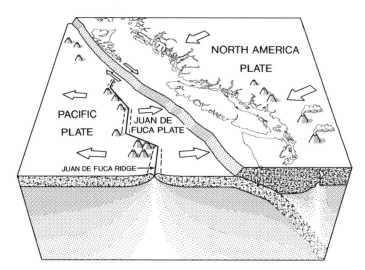

Today North America is moving westward away from the Mid-Atlantic Ridge at a rate of about two centimetres per year. The North American Plate is but one of several crustal plates which are moving either away, toward or sliding past one another. Currently, off the west coast of Vancouver Island, the North American continental plate is converging westward upon the eastward-spreading Juan de Fuca oceanic plate. Along the contact between the two plates the latter is descending, or being **subducted**, beneath the continent. This process of subduction of sea floor along the western margin of North America has been going on for the past 180 million years, ever since Pangea broke up. Now you may ask, 'Has this had anything to do with the formation of the Rockies?' Indeed it has.

WHO HAS THE RIGHT-OF-WAY?

I now suggest that you change your vantage point from Banff Avenue to the future site of the "Big Hill" above the town of Cochrane, west of Calgary. The time is about 170 million years ago during the middle part of the Jurassic Period. From behind you, from the east, streams are draining the continent and carrying their sediment loads to the western sea covering the flank of the miogeocline where they accumulate as the lower part of the Fernie Formation. The edge of the continent lies not far to the west of a line joining the modern locations of Penticton, Kelowna, Prince George and points north. To the west is open ocean as far as you can see. As you watch during the next million years or so you notice the gradual appearance of a vast tract of crust composed of multiple chains of volcanic islands, some surrounded by coral **atoll** reefs, which seems to be moving northward into the path of the westward drifting continent. Disregarding the rules of marine traffic (the starboard boat, i.e. North America, has the right of way) the stranger crashes into and is thrust up onto the continent.

Let me pause for a moment and explain the origin of the stranger. The piece of crust that crashed into North America is called a **terrane**. More specifically, it's called the **Intermontane Superterrane**. Terranes are fragments of the earth's crust, each of which preserves a unique geological history, different from those of neighbouring terranes. The boundaries between terranes are faults. For example, the Indo-Australian Plate, including the continent of Australia and part of New Guinea, is currently colliding with the Indonesian archipelago, a part of the Eurasian Plate. The collision zone is along a complex series of faults. The Indo-Australian Plate is largely composed of old continental rocks, much like those of North America, whereas the Indonesian

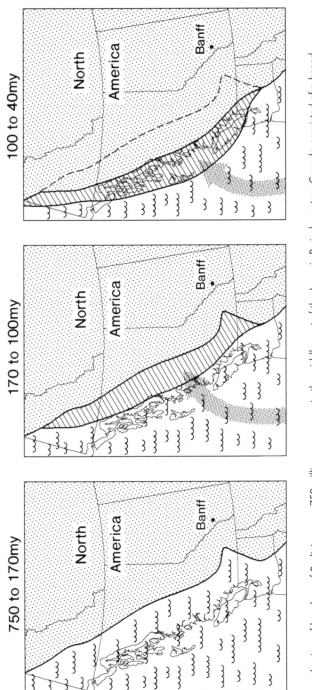

From the time of break-up of Rodinia, some 750 million years ago, to the middle part of the Jurassic Period, western Canada consisted of a broad continental shelf (miogeocline) covered by shallow tropical seas in which carbonate strata accumulated upon the shelf (left). About 170 million years ago the westward moving continent collided with a large fragment of northward-moving crust and mantle called the Intermontane Superterrane, itself an amalgamation of several smaller terranes (middle). Approximately 70 million years later the continent collided with the Insular Superterrane (right), following which, begining about 40 million years ago, several smaller terranes were added to Vancouver Island and western Alaska. Natural Resources Canada, *Of Rocks, Mountains and Jasper* by Chris Yorath and Ben Gadd. Dundurn Press. Reproduced with the permission of the Minister of Supply and Services Canada, 1995.

archipelago is dominated by comparatively young, oceanic volcanic and sedimentary rocks, including coral reefs. These two terranes have totally different geological histories, are composed of vastly different types of rocks and are currently colliding as a consequence of plate tectonic processes. What happened about 170 million years ago along the edge of North America is very similar to what is happening today in the southwest Pacific.

The Intermontane Superterrane consists of several smaller terranes which amalgamated prior to their collision with North America. Some consisted of chains of volcanic islands, called **island arcs** because of their arcuate shape, whereas others consisted of pieces of deep seafloor sedimentary and volcanic rocks, the latter formed at seafloor spreading ridges. Each of these smaller terranes has been given a name such as "Stikinia," "Quesnellia," "Slide Mountain"and"Cache Creek." These four terranes, and a few others, amalgamated to form the Intermontane Superterrane sometime during the Late Triassic Period, about 220 million years ago, and then crashed into North America some 50 million years later.

The effects of this first of two main collisions were dramatic, to say the least. Along much of its length, the edge of the Intermontane Superterrane, which today makes up most of the interior of British Columbia, was thrust up onto the western margin of the continent. As a consequence the continental margin was depressed by as much as twenty kilometres, where great heat and pressure were applied to its rocks. The effect was to cause the rocks to be crushed, intensely deformed, metamorphosed and intruded by masses of molten **magma** formed deep beneath the crust. The result was the formation of a great welt of traumatized rock extending along the full length of the collision zone, called the **Omineca Belt**, and which includes the Purcell, Selwyn, Cariboo, Omineca, Cassiar and Pelly mountains. And then it happened again.

By about 100 million years ago, during mid-Cretaceous time, another superterrane, called the **Insular Superterrane,** which included Vancouver Island, the Queen Charlotte Islands and parts of southeastern Alaska, had crashed into the edge of the Intermontane Superterrane. The effects of this second collision were no less dramatic than the first. Ray Price of Queen's University describes how the great blanket of strata which had accumulated on the miogeocline since the break-up of Rodinia, some 580 million years earlier, was sheared away from its Precambrian granitic basement, sliced up along a series of fractures, or **thrust faults**, and shoved eastward like a stack of shingles, such that the entire mass moved further inland by as much

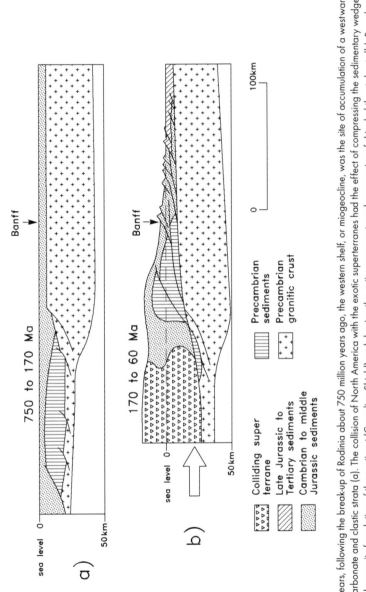

For some 580 million years, following the break-up of Rodinia about 750 million years ago, the western shelf, or miogeocline, was the site of accumulation of a westward thickening blanket of carbonate and clastic strata (a). The collision of North America with the exotic superterranes had the effect of compressing the sedimentary wedge, breaking it loose from the granite foundation of the continent (Canadian Shield) and shoving the entire mass eastwards as a series of shingled thrust sheets (b). Based upon a drawing by R.A. Price. Natural Resources Canada, *Of Rocks, Mountains and Jasper* by Chris Yorath and Ben Gadd. Dundurn Press. Reproduced with the permission of the Minister of Supply and Services Canada, 1995. Ma = millions of years ago.

as 250 kilometres. That's right; the rocks you see forming Mount Rundle and Cascade Mountain originally accumulated many tens of kilometres to the west.

A Tidal Wave of Rock

Let's return to our lookout above Cochrane. As just described, the effect of the two collisions was to cause the miogeocline to become subjected to intense compressive forces. As a consequence, the blanket of miogeoclinal strata, many kilometres thick, broke loose from its basement and became sliced up by thrust faults along which the rocks above these faults rose up and over those beneath. Like a giant advancing wave the ground surface rose above successive thrust faults, which progressively developed from west to east over a period of some 60 million years. In front of the advancing ground swell a trough developed, formed in response to the great load being placed upon the crust, which received enormous volumes of sediment eroded from the easterly advancing ground wave. Ultimately these sediments also became fractured by thrust faults until, about 60 million years ago, the pushing and shoving ceased. An excellent place to see these clastic strata is along the north side of the Trans-Canada Highway close to its intersection with the eastern entrance to the town of Banff. There they are assigned to the upper part of the Fernie Formation and Kootenay Group and consist of alternating beds of sandstone and shale, some of which contain large fossilized tree stumps. The part of the trough which includes the Fernie and Kootenay strata in the Canmore to Cascade Mountain area is called the Cascade Coal Basin (*see page 79*).

The scene that now confronts you is a broad highland, which, in the far distance at the location of Field, stands as much as 2,500 to 3,000 metres above sea level. Even farther west, in the Omineca Belt in the vicinity of Revelstoke, is a high plateau with a surface elevation of perhaps as much as six kilometres. Dinosaurs, such as *Triceratops* and various tyrannosaurs which had lived along the shores of the marine trough, had just been annihilated as a consequence of the effects of impact of a large meteorite in the region of the Yucatan Peninsula of Mexico. Rivers draining the plateau and highlands have carved V-shaped valleys between the thrusted sheets of carbonate, each of which forms a northwesterly trending longitudinal ridge. In some places rivers such as the Peace, and forerunners to the Athabasca, Saskatchewan and Bow rivers have managed to cut **antecedent** valleys *through* these ridges as they were being uplifted. The low hills in front of the upland, formed from sediments dumped into the trough

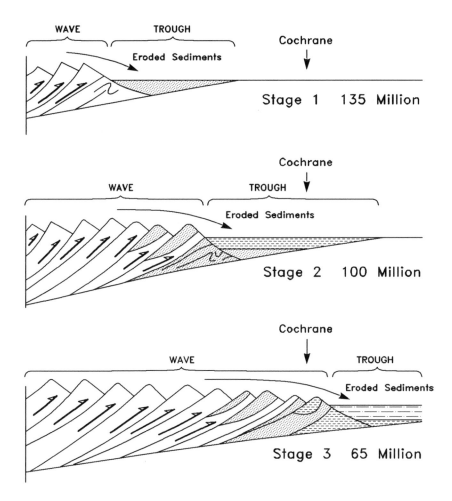

Like a gigantic eastward-moving wave, the Rockies were elevated along thrust faults. In front of the wave, a trough formed which received detritus that had been eroded from the rising mountains (Stage 1). As thrusting continued, the eroded sediments in the trough also became incorporated into thrust sheets (Stages 2 and 3). This process continued until thrusting ceased about 60 million years ago, following which the mountain front has been eroded back to its present position. Based upon a drawing by Ben Gadd, 1995.

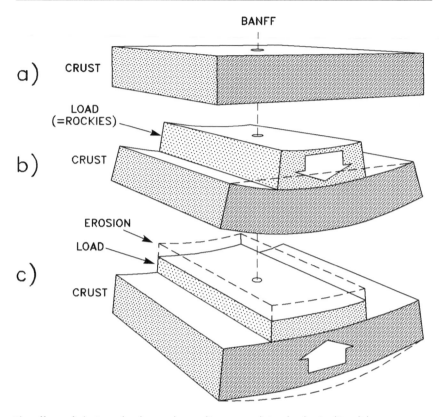

The effects of placing a load upon the earth's crust explain why the Rockies did not grow beyond a limiting height. Prior to the formation of the Rockies, the crust beneath the parks was a thick slab floating upon the denser, plastic mantle (a). When the Rockies (= load) were suddenly shoved eastward and placed upon the crust at the location of Banff the slab bent downwards under the load (b). As the forces of erosion wore the Rockies down (c) the load on the crust was reduced, allowing the crust to begin to return to its former position. By this means the maximum height of the Rockies has been maintained at a more or less constant elevation over the past several million years, during which the rebound of the crust has kept pace with the rate of erosion. Natural Resources Canada, *Of Rocks, Mountains and Jasper* by Chris Yorath and Ben Gadd. Dundurn Press. Reproduced with the permission of the Minister of Supply and Services Canada, 1995.

as a series of coalescing deltas, alluvial fans, beaches and bars today are seen as the Foothills of the Rockies.

For some fifty-eight million years the distant plateau and highlands were eroded by streams, rain and other processes. From above the peaks at Field, perhaps as much as five kilometres of strata were removed while in some places within the Omineca Belt, twenty kilometres of rock are estimated to have been stripped away. Does this mean that the mountains at Field and those of the Omineca Belt respectively were once five and twenty kilometres higher than they are

The characteristic geological signature of the Foothills of the Rockies is illustrated by this photo of Ram Falls on the Ram River, a tributary of the North Saskatchewan, south of Nordegg, Alberta. The cataract is formed from resistant sandstone of the Cardium Formation which, together with shales above and below, are part of the Cretaceous Alberta Group. The easternmost ridge of the Front Ranges appears in the far distance. (Photo by B. Hart.)

today ? No it doesn't.

It works this way. When a load of rock, such as the stacked up thrust sheets of the Rockies, is suddenly placed upon the crust, the crust subsides beneath the weight. As the rate of mountain building declines, at some point the rate of erosion will exceed the rate of mountain growth such that the crust will begin to rebound back toward its previous position as more and more of the applied load is stripped away. The tendency for the crust to behave in this way is called **isostasy** and by this means the Rockies have more or less been able to maintain their elevation even though a great deal of rock has been removed from them. The condition of isostasy, together with the amount of heat flowing upward from the lower crust and mantle, determines the elevation of the land surface. To reduce the surface elevation of a mountainous region by one kilometre, approximately five kilometres of rock must be removed by erosion. Insofar as twenty kilometres of rock are estimated to have been removed from the Omineca Belt, the implication is that at one time the plateau of the Omineca Belt was four kilometres higher than it is today, approximately six kilometres above sea level, or about the same height as the Tibetan Plateau. The same arguments suggest that the Rockies at Field were perhaps one kilometre higher than at present and those at the town of Banff about half a kilometre higher.

Where has all this rock gone? Apart from that which was removed by glaciation, a subject we are coming to soon, much of this material was carried out into the frontal trough by river systems draining eastward from the uplifted plateau and highlands; some, however, may have been carried westward to ultimately accumulate on the continental shelf off the west coast. It appears that much of the trough material accumulated in three major pulses, the first two apparently coinciding with the two collisions and the third with later additional northward displacements of the superterranes along an array of vertical faults called **strike-slip faults**. Each event caused the formation of broad coalescing **alluvial fans**, beaches, off-shore bars and many other types of sedimentary accumulations; these developed along the shores of the trough which, at times, enclosed a marine seaway extending from the Arctic Ocean to the Gulf of Mexico. The upper part of the Fernie Formation, the Kootenay Group, the Blairmore Group and the Alberta Group as well as the succeeding Brazeau and Paskapoo/ Porcupine Hills formations — all essentially composed of various combinations and amounts of shale, sandstone and conglomerate — form the strata exposed in the Foothills of the Rockies. The boundary between the Brazeau Formation of latest Cretaceous age and the Paskapoo and Porcupine Hills formations of earliest Tertiary age approximately coincides with the time when the dinosaurs were catastrophically obliterated, some 66 million years ago.

THE GLEAMING SHROUD

The last chapter in the formation of the Rockies of Banff and Yoho national parks has to do with glaciation. Beginning some 2 million years ago, during the **Pleistocene Epoch** of the **Quaternary Period**, vast areas of northern hemisphere continents were covered by thick sheets of ice, perhaps as many as eight times. The causes of these as well as earlier glaciations during the Precambrian Eon and Paleozoic Era, remain uncertain but are thought to be due to changes in climate induced by changes in ocean/atmosphere circulation patterns, which in turn are related to changes in the arrangement and average elevation of continents. Ultimately, however, the most important cause was variations in the amount of radiation received from the sun as a consequence of changes in the earth's rotational and orbital characteristics, the so-called **Milankovitch cycles** (*see Glossary*).

The record of the older glaciations that affected North America

The large cirque on the east side of Castle Mountain encloses Rockbound Lake, a tarn, its basin quarried by glacial ice.

during the Pleistocene Epoch is virtually absent in the region of the two parks; only the last, or **Wisconsinan** glacial record, is preserved. Beginning about 80,000 years ago valley glaciers spread outward from ice domes that had formed at high elevations in the Cordillera. During times of advance these valley glaciers widened and straightened the former trunk river valleys of the Athabasca, Saskatchewan and Bow rivers, and at times of retreat left widespread glacial outwash sands and gravels on the valley floors. The glacial carving gave many valleys a characteristic **U-shape**, such as that of the Spray River west of Banff (*see photo on page 70*). In many places, high along the sides of these glacially widened valleys, smaller tributary valleys were left as **hanging valleys** to pour their waters into the main valleys as cataracts such as those near the southeastern end of Lake O'Hara in Yoho National Park (*see page 97*). Above the valleys the ice carved the many peak shapes we see today: **horn peaks** like Mount Assiniboine in Mount Assiniboine Provincial Park and Mount Chephren along the Icefields Parkway (*see photo on page 67*), formed by the headward erosion of three or more glaciers carving a central spire; **aretes**, steep narrow ridges caused by glaciers gnawing away on both sides, such as the ridge connecting Narao Peak, Popes Peak and Mount Victoria along the Continental Divide above Lake Louise; and **cirques**, amphitheatre-like basins quarried by glaciers and enclosing small lakes called **tarns**, such as the large cirque on the east flank of Castle Mountain enclosing Rockbound Lake. So it is to be remembered that the actual shapes of the mountains and valleys you see are due to erosion, particularly that of glacial ice. The mountains are the remnants of that erosion.

The deposits left behind by melting and retreating ice also provide many landforms. **Lateral** and **terminal moraines** are seen as ridges of gravel along the sides and fronts of retreating glaciers. The processes currently forming these and several other depositional features are visible at the Columbia Icefields in Jasper National Park. The interested reader is referred to a book similar to this one, written by myself and Ben Gadd, entitled *Of Rocks, Mountains and Jasper – A Visitor's Guide to the Geology of Jasper National Park* (*see Sources and Additional Reading*).

Our knowledge of the history of glaciation in Canada's mountain national parks is due largely to the efforts of Nat Rutter and his students at the University of Alberta and Lionel Jackson of the Geological Survey of Canada. That history is one of several advances and retreats of Cordilleran ice which reached a maximum elevation of at least 2,440 metres above sea level in the Banff area. The **Laurentide Ice Sheet**, which had spread westwards from centres in northern and eastern Canada to cover most of the northern part of North America, is thought to have advanced into the valleys of the Foothills at various times. The interaction of Cordilleran and Laurentide ice beyond the mountain front is unknown in several areas, however, one such interaction left spectacular results in many places in central and southern Alberta. Tongues of ice flowing out of the Rockies met the Laurentide Ice Sheet then flowed southward, forming a line of contact between the two glaciers. Some of these ice tongues carried house-sized blocks of quartzite upon their surfaces, masses of rock that had fallen onto the ice surface from over-steepened, ice-sculpted cliffs in the mountains. As the ice moved out to meet and flow alongside the Laurentide sheet, these blocks tended to align themselves along the contact. When the ice melted and retreated, the blocks were let down upon the surface of the plains where they now form a chain of **erratics** in western Alberta. One of these erratics is seen at Okatoaks, south of Calgary, where it is known as "The Big Rock."

The history of glaciation in the Bow Valley during the last 24,000 years of the Wisconsinan glacial epoch is recorded in deposits left by advances and retreats of the ice. Glacial **till**, composed of an unsorted mixture of pebbles enclosed within dense clay and silt, is believed to have been deposited by advancing ice during the so-called Bow Valley Advance which extended to well beyond the mountain front. The ice then retreated to about the location of Banff leaving in its wake an extensive train of gravel and sand outwash. The subsequent Canmore Advance saw the ice front reach as far as the mouth of Kananaskis River, followed by its retreat to west of Castle Junction. After a brief readvance, ice left the Bow Valley about 9,300 years ago.

An interesting example of the effects of glaciation upon river drainage systems can be seen in the vicinity of Banff townsite and is best viewed from Sulphur Mountain's upper gondola terminal (*see photo on page 67*). Prior to the establishment of its present course between Tunnel Mountain and Mount Rundle the Bow River flowed eastward between Tunnel and Stony Squaw mountains; although, at least for a time, during retreat of the Canmore ice tongue, the principal drainage appears to have been eastward via the Cascade and Lake Minnewanka valleys. At that time glacial ice occupied all of the mountain valleys up to an elevation of about 2,440 metres. Glacial meltwater streams flowing northward on or alongside the glacier in the Spray River valley probably cut the gap separating Tunnel Mountain and Mount Rundle during deglaciation. The Bow River was probably then diverted to its present course through this gap, either by ice remaining between Tunnel and Stony Squaw mountains or by thick accumulations of outwash deposited by the melting ice. The diversion resulted in the formation of Bow Falls near the Banff Springs Hotel where the Bow River cascades over Triassic siltstone and shale of the Sulphur Mountain Formation.

WHAT NOW?

Throughout the past 9,000 years or so the present physiography of the mountains was established through further erosion of the glacial landscape. The Bow and North Saskatchewan rivers broadened their valleys, which, beyond the mountain front, are characterized by spectacular river terraces, particularly those of the Bow in the vicinity of Cochrane. Locally, within the mountains, erosion produced dramatic effects such as at Johnson Canyon where, according to Parks Canada, some 8,000 years ago Johnson Creek was diverted by a landslide from off Mount Ishbel, which brought masses of Livingstone Formation carbonates into the valley floor. Since that time a 200-metre-deep canyon has been cut. Other landslides include those visible on the west-facing slope of Castle Mountain where large blocks of Cambrian Cathedral and Eldon carbonates collapsed from the main edifice and

The steep slopes of loose scree beneath the upper carbonate cliffs of Cathedral Mountain are a source of debris flows which can threaten the CPR railbed and the Trans-Canada Highway alongside the Kicking Horse River. These debris flows are induced by sudden releases of water from Cathedral Glacier, just visible to the right of the peak.

came to rest upon Precambrian Miette Group slates forming the mountain's lower slopes (*see page 88*). Modern landslides and debris flows are a continuing problem in the Kicking Horse River valley near Field. From high on Cathedral Mountain the Cathedral Glacier commonly will suddenly release between 10,000 and 24,000 cubic metres of water, either from a small lake on the south side of the glacier or from large volumes of water entrained in the ice. These so-called **jokulhlaups**, an Icelandic word for sudden releases of glacier-entrained water, typically result in mobilizations of approximately 100,000 cubic metres of loose talus on the steep slopes of the mountain. Since 1925 several such events have damaged both the highway and the C.P.R. mainline. With such activity occurring frequently throughout the past several thousand years, the valley has become filled with this debris across which the Kicking Horse River braids the upper part of its course; the lower part of the valley is a deep canyon which probably formed through headward erosion of the river since the end of glaciation (*see page 104*).

It is particularly in the Front Ranges, where resistant carbonate and recessive shaly formations alternate with one another, that one can appreciate the relationship between the effects of mechanical ero-

Hole-in-the-Wall is a cave excavated into steeply inclined carbonate strata of the Livingstone Formation on the southwest side of Sawback Range. The cave may have been formed through a combination of frost-shattering and dissolution by a stream flowing along the flank of a Bow valley-filling glacier.

sion and altitude. By mechanical erosion I mean the effects of daily freezing and thawing of water in the cracks and fractures of rocks, as opposed to chemical erosion, more prevalent in tropical to subtropical climates, where carbonates are more readily dissolved away. The freezing of water on bedding surfaces and in fractures and pores of rocks results in their expansion and consequent fragmentation. At temperate and arctic latitudes where mechanical erosion processes are more significant, the effects of these generally increase with elevation up to the point where water, entrained in fractures, is frozen the year round. In the Front Ranges, however, mountain-top elevations are not so high, thus the fragmentation caused by the daily freezing and thawing of fracture-entrained water, coupled with the effects of gravity, combine to further erode the old glacial landscape. The less resistant shaly formations erode more readily than the carbonates, the differential effect being greater at higher elevations. In this way the mountains are continuously being eroded, as evidenced by the talus fans which mantle the gentler slopes beneath the cliff-forming carbonates, the latter commonly spalling off large blocks as the shaly rocks are removed beneath them. It is this process which has resulted in the retreat of the mountain scarps, particularly those at the mountain front, ever since the retreat of Cordilleran and Laurentide ice some

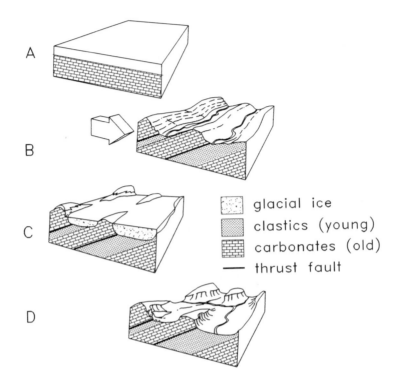

Development of the present physiography of the Rockies began with the accumulation of carbonate and clastic strata during some 580 million years (a). As a consequence of collisions between North America and far-travelled oceanic terranes, these strata were crumpled, broken into thrust sheets and shoved eastwards. Erosion by rivers and wind cut into the thrust sheets, exposing carbonate layers forming the ridges while the softer shales eroded more deeply to form the intervening valleys (b). During the past 2 million years glaciers have further eroded the landscape (c), leaving the shapes of the mountains and valleys we see today (d). Natural Resources Canada, *Of Rocks, Mountains and Jasper* by Chris Yorath and Ben Gadd. Dundurn Press. Reproduced with the permission of the Minister of Supply and Services Canada, 1995.

10,000 years ago.

Another effect of alpine erosion processes is displayed by Hole-In-The-Wall, a cave clearly visible on the southwest flank of Sawback Range, some 480 metres above the Trans-Canada Highway. Whereas most caves in the Rockies were formed through underground stream dissolution of either flat-lying or gently inclined carbonate strata, Hole-In-The-Wall has been formed in very steeply inclined strata and thus is not easily explained by normal **karst topography**-forming processes. According to Derek Ford of McMaster University, an expert in Rocky Mountain cave systems, the cave is about 30 metres deep, is filled with loose sediment of probable glacial origin and displays the effects of frost-shattering on its walls and nearly horizontal roof. Derek suggests that frost action has substantially enlarged Hole-In-The-Wall, and moreover, that it may be entirely of frost-weathering origin, a so-called "frost pocket" and very common in alpine carbonate regions. Alternatively, he suggests that the cave might have originated through dissolution by streams flowing along the flank of a glacier that filled the Bow Valley, and that it has since been enlarged by frost action.

Following retreat of the ice, the land surface rebounded back to the level it was at before the load of ice was placed upon it. During this period of **glacial rebound**, a consequence of isostasy (*see page 38*), many mountain rivers deepened their valleys and carried the eroded debris out into the Foothills. As the land surface slowly rose, the Bow River shifted its course many times across a broad valley which, in the Foothills, displays several terraces, each cut by the river as it strived to reach its **base level**, below which it cannot erode any deeper. During this time much of the land surface was covered with fine silt and clay which was swept by winds moving through the mountain valleys. This wind-driven dust, or loess, accumulated in several places. Such deposits near Saskatchewan Crossing contain two volcanic ash layers: the lower originated from the eruption of Mount Mazama, which created Crater Lake in Oregon 6,670 years ago; and the upper from an eruption of Meagre Mountain north of Vancouver about 2,500 years ago.

The Mountains We See Today

Whenever you are on a commercial airline flight across the Canadian Cordillera, the Rockies stand out as distinctly different from the other mountain systems. Throughout British Columbia, much is rolling plateau terrain bordered by seemingly disorganized mountains. The Coast Mountains, for example, consist of extraordinarily rugged masses of granitic rocks within which you can rarely see any discernable continuity in form, shape or structure from one mountain to another. Much the same can be said for the **Columbia Mountains**, which include the Selkirk, Purcell and Cariboo mountains, although in these cases individual mountains commonly show recognizable structure. When you cross eastward above the **Rocky Mountain Trench**, part of a system of deep, linear, fault-controlled valleys extending the length of the Canadian Cordillera (*see page 109*), the Rockies are instantly recognizable because of their continuity in form, shape and structure, both along each range and from range to range. The ranges are disposed in parallel, linear ridges extending in a southeast to northwest direction. The ridges are commonly narrow with steep to cliffy northeast facing slopes and less steep westerly inclined slopes. From range to range, each separated by comparatively wide valleys, this characteristic

The view to the southeast from Abraham Lake along the David Thompson Highway (Hwy. 11) shows the profile of the low, rolling Foothills, composed of Mesozoic clastic strata, adjacent to the prominent Paleozoic carbonates of the Front Ranges.

Ray Price took this photograph of typical front range structural and physiographic styles. The shot, looking to the northwest, is taken from above the Trans-Canada Highway near Sundance Canyon. On the left is the Borgeau Thrust Sheet (Sundance and Sawback ranges), in the middle the Sulphur Mountain Thrust Sheet (Sulphur Mountain and Vermillion Range) and on the right the Rundle Thrust Sheet (Mount Rundle and Cascade Mountain). In the upper right is the Palliser Range. Each thrust sheet consists of Paleozoic, dominantly carbonate strata, which have been thrust over Mesozoic clastic rocks underlying the valleys. The surface traces of the Bourgeau, Sulphur Mountain and Rundle thrust faults occur just below the steep, east-facing cliffs of each range. Each of these thrust faults consists of a westerly inclined fracture along which the thrust sheet has moved from west to east. In this way the thrust sheets are tilted slabs, resembling shingles on a roof, in which the same formations are repeated from range to range.

landform style is repeated to where the mountains are bordered by the low-standing Foothills. It matters not whether your flight across the Rockies is to or from Edmonton, Calgary or wherever, the pattern is the same.

When driving, you may notice that from west to east the colours change from dark, sombre hues to light grey tones. The rocks become increasingly well layered, or stratified, and for the most part, the inclination of the strata is the same as the western backslopes of the ranges. All of these features are direct consequences of the kinds of rocks from which the mountains are made and the kinds of structures resulting from the forces that elevated the rocks into mountains.

"THE BLUE CANADIAN ROCKIES"

In a general sense the Rockies may be divided longitudinally into three parts: the **Foothills**, **Front Ranges** and **Main Ranges**. The Front and Main ranges occur in Banff National Park, whereas only the Main Ranges are found in Yoho. Along the Trans-Canada Highway leading westward from Calgary the Foothills extend to the mountain front at Mount Yamnuska, the prominent cliff of Cambrian Eldon Formation dolomite standing proudly above Cretaceous shales of the Belly River Formation. The contact between the old rocks of the cliff and the younger rocks of the lower slopes is a thrust fault called the McConnell Fault, named after an early geologist/explorer of the Geological Survey of Canada. For those interested in the geological exploration history of Canada I recommend the scholarly book *Reading the Rocks*, by Morris Zaslow (*see Sources and Additional Reading*).

From the mountain front at Mount Yamnuska the Front Ranges consist of a series of parallel, northwesterly-trending ridges, or ranges, some fifty-three kilometres in width. The most characteristic feature of these ranges is their prominent cliffs of Devonian and Carboniferous carbonate, which, in the mid-day summer sun reflect the pale blue-grey tone that inspired Wilf Carter's nasal crooning of "In the Blue Canadian Rockies." Between the ranges, softer and more easily erodible rocks form the valley floors such that from range to range, hard, comparatively erosion-resistant carbonates rise above less erosion-resistant clastics. The underlying cause of this repetitious character of the Front Ranges has to do with thrust faults.

Faults are marvelous things. People have them, so do mountains. In fact, were it not for faults, our Rockies, like most people, would be pretty dull. The geological recognition of thrust faults in the Rockies goes back to the last century; however, we owe our current understanding of these structures largely to Ray Price, of Queen's University. Together with colleagues such as Eric Mountjoy of McGill University and their students, Ray has characterized the Rockies as a huge mass of rocks that is internally delaminated by a myriad of westerly inclined fractures which merge downwards with one another and with the surface of the underlying granitic Precambrian basement. Along each of these fractures, the rocks above have moved upwards and eastward over those beneath such that along each fault the strata above the fault are older than those beneath. Older rocks thrust over younger strata is the characteristic **structural style** of the Rockies. In this way the several carbonate formations of the Front Ranges are repeated, range after range (*see page 59*).

A good illustration of this structural style can be seen from the Vermilion Lakes viewpoint along the Trans-Canada Highway just west of the westernmost entrance to Banff townsite. There you can see two mountains: Mount Rundle in the Rundle Range on the left, and Sulphur Mountain in the Sulphur Mountain and Goat Range to the right. In each range the strata are tilted downward to the west at between 50 and 60 degrees and, in each mountain the same group of three formations is repeated, resulting in the classical mountain form that I talked about earlier: a lower, east-facing cliff composed of Devonian carbonates of the Palliser Formation, a middle slope of limy shale and shaly limestone of the Banff Formation and an upper east-facing cliff of Livingstone Formation carbonates, the latter two formations of Carboniferous age. The western slopes of each range are inclined at approximately the same angle as that of the strata. In the lower tree-covered slopes of each range are strata assigned to the Devonian Alexo Formation and Fairholme Group, beneath which is the **surface trace**, or fault line, of a thrust fault, the Rundle Thrust on Mount Rundle and the Sulphur Mountain Thrust on Sulphur Mountain. Each fault is gently inclined, or **dips**, to the west such that the older, Devonian, mainly carbonate strata have been shoved eastward and overtop of younger Mesozoic clastic rocks which occur in the valleys of the Bow and Spray Rivers. Farther west the Sundance Range contains the same group of formations again, but there Cambrian rocks also are exposed beneath Devonian strata above the Borgeau Thrust. The surface traces of each of these thrust faults — the Rundle Thrust, the Sulphur Mountain Thrust and the Bourgeau Thrust — can be traced for many kilometres along the trends of these ranges.

So it is with all the other thrust faults of the Rockies. In this way the several ranges of the Rockies are like cedar shingles on a roof. The rocks above each thrust fault form a slab of strata which geologists

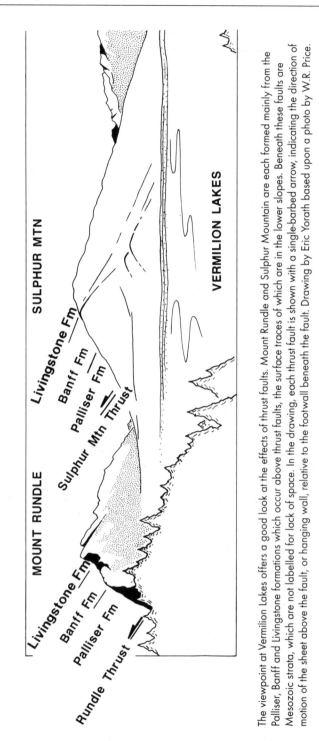

The viewpoint at Vermilion Lakes offers a good look at the effects of thrust faults. Mount Rundle and Sulphur Mountain are each formed mainly from the Palliser, Banff and Livingstone formations which occur above thrust faults, the surface traces of which are in the lower slopes. Beneath these faults are Mesozoic strata, which are not labelled for lack of space. In the drawing, each thrust fault is shown with a single-barbed arrow, indicating the direction of motion of the sheet above the fault, or hanging wall, relative to the footwall beneath the fault. Drawing by Eric Yorath based upon a photo by W.R. Price.

Devonian carbonate and limy shales form the peaks of the Three Sisters above the town of Canmore, Alberta. The leftmost of the three is carved from strata of the Palliser Formation which are folded into a syncline lying on its side above the Rundle Thrust, the surface trace of which is at the base of the lowermost cliff above the trees. The middle and right peaks together consist of a westerly inclined succession of Palliser, Banff and Livingstone formations strata.

call **thrust sheets**. Each thrust sheet forming a range has been carved by glacial ice and other erosional processes into individual peaks.

As you drive through the Front Ranges, from the mountain front just west of Seebe beneath Mount Yamnuska to their boundary with the Main Ranges at Castle Mountain, the Trans-Canada Highway follows the valley of the meandering Bow River, twice changing course from a southwesterly to a northwesterly direction. As the highway crosses through the eastern Front Ranges you get a view of the thrust sheets in cross-section from Mount Yamnuska to the Fairholme Range at Gap Lake. From there the highway and river valley turn northwesterly to traverse between the ranges, or, as geologists would say, "along **strike**." On this part of the route, beneath the peaks forming the Three Sisters and the massive northeastern face of Mount Rundle, the geology changes little until the highway again turns southwesterly to cut through the Rundle, Vermilion and Sawback/Sundance Ranges, where the internal structure of individual thrust sheets is once more exposed. Shortly past the Vermilion Lakes viewpoint the highway again turns northwesterly to travel obliquely across Sawback Range; here it passes beneath the southeastern snout of the splendid Cambrian carbonate

From a viewpoint close to Golden, strata of the Ordovician McKay Group have been folded into an anticline that can be seen on the south wall of Kicking Horse canyon. Note that the anticline is asymmetrical, or tilted, to the southwest. This is in contrast with folds in the eastern Main and Front ranges which are commonly tilted towards the east (i.e., the Three Sisters).

ramparts of Castle Mountain which, at this latitude, is the eastern-most mountain of the Main Ranges. Castle Mountain sits above the Castle Mountain Thrust, which is one of many subsidiary faults, or **splays**, of the Simpson Pass Thrust. The Simpson Pass is one of the longest thrust faults in the Rockies, extending from south of Lake Louise to north of Jasper, a distance of some 360 kilometres. From Castle Junction, the intersection of the Trans-Canada Highway and Highway 93, to just past Lake Louise, the highway and river valley trend again along strike, but now within the Main Ranges. The course of the highway here and along most of the Icefields Parkway from Lake Louise to Jasper follows the surface trace of the Simpson Pass Thrust.

Although thrust faults are very important to the structural style of the Main Ranges, **folds** in the forms of **anticlines** (upfolds) and **synclines** (downfolds) are common, much more so than in the Front Ranges. The reason for this is that when the rocks were being stressed by the forces of collision, the sheets of carbonate in the Front Ranges behaved like brittle concrete — they tended to break into great slabs rather than bend as did the weaker, or less **competent**, limy clastic rocks of the Main Ranges. As you pass southwestward through the

Main Ranges, the ranges become less well ordered, less predictable in form and shape. Instead of the pale, blue-grey colour of the Front Ranges, the rocks of the Main Ranges display dark, sombre tones. Structures here are very complex, consisting of folded faults, huge upright, fan-like multiple fold structures and intensely compressed rocks in which the shaly strata of the Cambrian Chancellor Formation have been strained to the point at which closely spaced microscopic fractures, or **cleavage**, have utterly disrupted and obliterated the original stratification.

Continuing on past Field in Yoho National Park you pass through several large folds while following the valley of the Kicking Horse River. The upper Kicking Horse valley is broad, open, glaciated, and debris-filled; however, you soon enter the narrow, twisting canyon of the river where roadside outcrops consist of contorted greenish-grey and yellowish-grey limy clastic strata of the Ordovician McKay Group and black silty shale of the Glenogle Formation. At a viewpoint above the town of Golden you can see an inclined fold in McKay Group strata in the cliffs on the south wall of the canyon (*see page 54*).

The western boundary of the Main Ranges and the Rocky Mountains is the **Southern Rocky Mountain Trench**, on the valley floor of which is located the town of Golden. On the west side of the trench is the Dogtooth Range of the Purcell Mountains, part of the Omineca Belt. The Southern Rocky Mountain Trench was formed by **normal faults** whereby the rocks underlying the floor of the valley were dropped downwards relative to those on the valley walls. Normal faults are steeply inclined to vertical fractures caused by tension in the crust; as opposed to thrust faults which are gently inclined and caused by compression.

There are several other normal faults, particularly in the Main Ranges (*see page 58*), which are believed to have formed after the pushing and shoving had ceased. As if uttering a great sigh of relief when it was over, the earth's crust, elevated to highlands and plateaus, relaxed such that some elevated blocks collapsed downward along these faults. Whereas thrust faults are fractures along which the thrust sheets moved as much as 250 kilometres, normal faults most commonly show smaller displacements of tens to thousands of metres. Between Lake Louise and Field you cross several normal faults which disrupt Cambrian strata of the Gog Group as well as Precambrian strata of the Miette Group.

As we have seen, the modern landforms of the Rockies are due to the forces of erosion, including glaciation, which acted upon large tilted thrust sheets. From east to west across the Front Ranges, you

The combined effects of topography, climate and wind direction favour the accumulation of glaciers and icefields on the east-facing slopes of many ranges in the Rockies such as Snowbird Glacier on the east face of Mount Patterson along the Icefields Parkway.

pass through seven of these tilted slabs, each making up a range. The ranges are dominated by thick formations of Paleozoic carbonate whereas the intervening valleys are floored by softer Mesozoic clastics. This **structurally controlled** physiography, resulting in steep east-facing cliffs and more gently inclined western slopes from range to range, has been further enhanced by our modern climate and weather patterns. Due to the northern latitude of the Rockies, their northwest-southeast orientation and the southwesterly inclination of their strata, the southwesterly-facing slopes are warmest because they receive the most sunlight late in the day when the air temperature is warmest; in contrast the northeast-facing steeper slopes are comparatively cool, receiving the sun earlier in the day and for less time. West-facing slopes are thus likely to have the least snow accumulation in winter, whereas northeast-facing slopes tend to retain their snow, well into summer and commonly until the following winter. Additionally, the wind in this region is mainly from the west and tends to blow the snow up the western slopes and over the ridge crest where it further contributes to the accumulation on the northeast side. The net effect of these conditions is the maintenance and development of glaciers on the northeast sides of the ranges where they quarry amphitheatre-like cirques into the mountain slopes. Consequently, the crestlines along the northeast sides of the ranges are commonly serrated with cirques, some with remaining glaciers and tarns. An excellent place to observe this phenomenon is along the Icefields Parkway between Bow Lake and Saskatchewan Crossing where cirques and glaciers are common on the east-facing slopes but virtually absent on those facing to the west.

The regional physiography of the Rockies has resulted in its modern **trellis** drainage system. Due to the northwest-to-southeast alignment of the ranges, the geometry of stream courses is systematic and regular in form. On either side of the ridges small, **consequent** streams drain the dip slopes and carry their sediment loads to the main trunk, or **subsequent**, streams flowing along the floors of the valleys separating the ranges. These streams in turn empty into the main antecedent rivers, such as the Bow and North Saskatchewan rivers, that have cut through the ranges.

THE GEOLOGICAL MAP

The geological map shown on page 58 depicts the surface of the ground, stripped of its forests, soil and glacial sediment cover, with the rocks labelled and patterned according to their age and the **tectonic** environments that existed when they were deposited. The adjective "tectonic" refers to the general state of the crust in a given region and the kinds of forces acting upon it. In this case the map shows the many formations of the Rockies grouped into larger categories according to broad time intervals during which they accumulated and according to the tectonic circumstances that existed during that time interval. The pattern on the map identified as "uP" on the legend refers to those regions underlain by dominantly clastic strata of late Precambrian age (the Miette Group) which accumulated across the torn edge of the continent as a consequence of the break-up of the supercontinent Rodinia, some 750 million years ago (*see pages 12 to 17*). The pattern designated "CD" refers to sedimentary rocks of Cambrian to Devonian age which accumulated on the westerly sloping miogeocline and which change in facies from eastern, shallow-water carbonates to western, deep-water shales along the dashed line extending through Yoho National Park (*see pages 17 to 27*). The pattern labelled "DP" shows the distribution of Devonian, Carboniferous and Permian formations, which are dominantly shallow-water carbonate bank and reef strata and organic clastic deposits. These accumulated on the stable eastern reaches of the miogeocline and upon the broad, undeformed **platform** beneath the plains (*see pages 27 to 32*). The

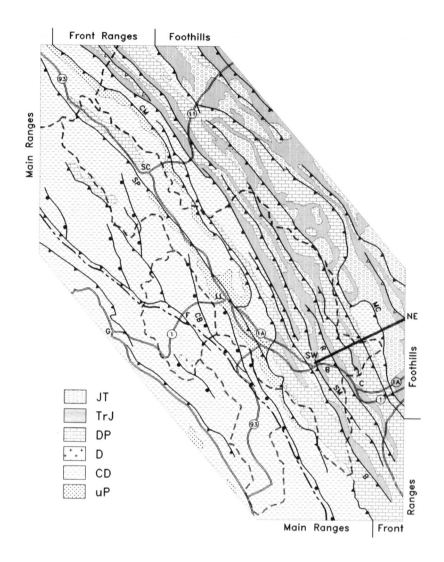

Geological map of Banff and Yoho national parks and adjacent areas. Screened dashed lines = park boundaries. Lines with triangles = the surface traces of thrust faults, triangles are drawn on the side which is up-thrust (hanging wall). Lines with circles = surface traces of normal faults, circles are drawn on the down-dropped side (hanging wall). uP = Precambrian strata of the Miette Group. CD = Rocks of Cambrian to Middle Devonian age. D = igneous rocks of Devonian age in southern Yoho National Park. DC = strata of Upper Devonian to Permian age. TrJ = rocks of Triassic to Middle Jurassic age. JT = strata of Upper Jurassic to Tertiary age. Dashed line = locus of facies change in Lower Paleozoic rocks. Mc = McConnell Thrust, R = Rundle Thrust, SM = Sulphur Mountain Thrust, B = Borgeau Thrust, CM = Castle Mountain Thrust, SP = Simpson Pass Thrust, CB = Cataract Brook normal fault. C = Canmore, B = Banff, LL = Lake Louise, SC = Saskatchewan Crossing, F = Field, G = Golden. The boundaries between the Foothills, Front Ranges and Main Ranges, all of which are thrust faults, are indicated around the periphery of the map. SW – NE = line of cross-section shown on page 59.

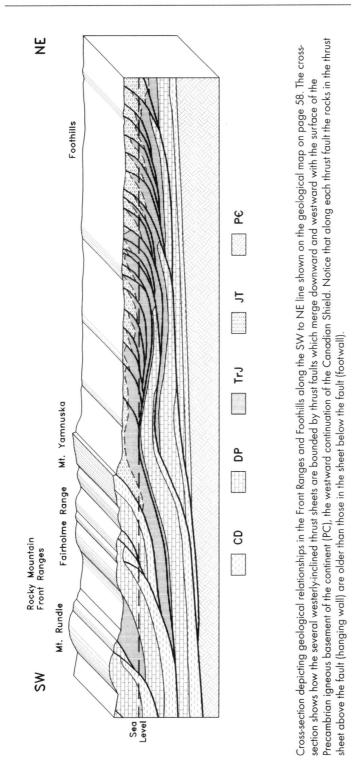

Cross-section depicting geological relationships in the Front Ranges and Foothills along the SW to NE line shown on the geological map on page 58. The cross-section shows how the several westerly-inclined thrust sheets are bounded by thrust faults which merge downward and westward with the surface of the Precambrian igneous basement of the continent (PC), the westward continuation of the Canadian Shield. Notice that along each thrust fault the rocks in the thrust sheet above the fault (hanging wall) are older than those in the sheet below the fault (footwall).

small area of crosses, labelled "D," in southern Yoho National Park, shows one of the very few localities of igneous rocks in the Rockies; these are of Devonian age and are named the Ice River Complex (*see page 58*). The pattern labelled "TrJ" shows the distribution of Triassic to Middle Jurassic carbonate and clastic rocks which accumulated on the miogeocline as a westward-thickening blanket of sediments. The clastic components of this succession of strata were entirely derived from eastern sources within the continent and include the lower part of the Fernie Formation of the Cascade Coal Basin (*see page 36*) and the Foothills (*see page 40*). The pattern designated "JT" represents those sediments of Late Jurassic to Tertiary age which accumulated in the trough in front of the rising and eastward-moving thrust sheets that formed as a consequence of the collisions of two large superterranes with the ancient western edge of the continent. These sediments were derived from erosion of those thrust sheets and from other western sources such as the Omineca Belt (*see pages 36 to 40*).

The orientations of the different patterned areas on the map arise from geological mapping of the distribution of the many rock forma-tions seen at the surface. From this we are able to interpret the geological structure beneath the ground as shown on the cross-sec-tion. The map is based upon one compiled by John Wheeler of the GSC which in turn was derived from more detailed maps published by the Geological Survey of Canada and authored by several people including Ray Price and Eric Mountjoy (*see Sources and Additional Information*).

The prominent striped pattern of the Front Ranges reflects the presence of thrust faults which have tilted thick thrust sheets or slabs of sedimentary strata that were originally horizontal. Notice how, from range to range, the patterns are repeated. The thrust faults are de-picted by heavy dark lines with teeth on them; the teeth point to the **hanging wall** of the fault, which is the slab or sheet above the fault. The slab beneath the fault is called the **footwall**. In the case of the Rockies, where thrust sheets are stacked one upon the other like shin-gles on a roof, the footwall of any given fault is the hanging wall of the next fault to the east.

The dark line labelled SW at its southwestern end and NE at its northeastern end depicts the position across the Front Ranges where the cross-section shown on page 59 is drawn. The cross-section de-picts the geometry of Front Ranges thrust sheets as they would appear in a vertical slice through the Rockies along the line shown on the map. On the cross-section you can see how these thrust faults have delaminated the originally horizontal strata, which then have been displaced along the faults such that the rocks above the faults are older than those beneath. Notice also that these thrust faults are believed to merge downward onto the surface of the deep Precambrian granitic basement. This merging of thrust faults to a common surface is called a **décollement**.

The change in pattern across the map, from the Front to the Main ranges, is quite dramatic. The striped pattern of the former becomes less pronounced, particularly in the western part of the Main Ranges. Also, the number of patterns reflecting the number of formations is much less in the Main Ranges where rocks of Precambrian to early Ordovician age are all that remain after some 100 million years of erosion. In the western Main Ranges the map shows many northwesterly-aligned normal faults and few thrust faults. Not shown are several large anticlines and synclines, a result of the change in facies from stress-resistant carbonates in the Front Ranges to weaker shaly rocks in the Main Ranges.

And so it was that over some 750 million years our Rockies were formed. In the ordered, pale, blue-grey cliffs of the Front Ranges, the emerald green stillness of mountain lakes and in the white starkness of snow, ice and sky, the Rockies of Banff and Yoho are among the lasting gems of this country. To know something of how they came to be and of what they are made is to better appreciate their grandeur and the silent depths of time.

PART TWO

In this part of the book I discuss the geology of specific localities in the two parks. For the most part they are situated along the main highways or are easily accessible by side roads to well known tourist destinations. Some described localities are outside the boundaries of the parks but are included for completeness and continuity. Those along the Trans-Canada Highway, or accessible from it, are described in order from east to west, beginning at the mountain front just east of Lac des Arcs. Others along the Icefields Parkway are presented from south to north. All localities are shown on the map on page 5.

The Trans-Canada Highway from the Mountain Front to the Rocky Mountain Trench

Mount Yamnuska and the Mountain Front

From about the intersection of the Trans-Canada Highway (Hwy 1) and the Kananaskis Highway (Hwy 40) you get an excellent view to the north of the mountain front, including Mount Yamnuska.

The imposing cliffs of Mount Yamnuska (an Assiniboine word for "sharp cliff" or "sheer-faced rock") are composed of carbonate rocks of the Cambrian Eldon Formation, about 535 million years old (*see photo on page 71*). The lower slopes of the mountain are formed from Cretaceous shale and siltstone strata of the Belly River Formation which are about 75 million years old. The contact between these two formations is the McConnell Thrust.

The McConnell Thrust, the surface trace of which is more than 200 kilometres in length, here separates the Foothills from the Front

The massive southeast face of Cascade Mountain illustrates the classic triplet of the front ranges: lower cliffs formed from Upper Devonian carbonates of the Palliser Formation; middle slopes shaped from limy shales and shaly carbonate of the Carboniferous Banff Formation and upper cliffs carved from Carboniferous carbonates of the Livingstone Formation. The surface trace of the Rundle Thrust occurs beneath the lower cliffs, at the base of Fairholme Group strata and above Mesozoic rocks seen in the middle distance beside the highway. (Photo by T.P. Lewis.)

From the road leading to Moraine Lake the mountains surrounding the Valley of the Ten Peaks form an impressive wall of Cambrian quartzite and carbonate. The Valley of the Ten Peaks owes its origin to an alpine glacier which widened the original V-shaped valley and which truncated two spurs extending southwesterly from Mount Temple.

Above the emerald-green waters of Lake Louise, the Victoria Glacier dresses Cambrian strata of the Gog Group.

Mount Wilson at Saskatchewan Crossing exposes four type sections of lower Paleozoic formations which are widespread throughout the Canadian Rockies. One of these is the Mount Wilson Formation which forms the uppermost prominent cliff of the clouded middle peak.

Mount Chephren on the Icefields Parkway is a classic example of a horn peak.

That Tunnel Mountain was once part of Mount Rundle is clearly evident when viewed from the upper terminal of the Sulphur Mountain Gondola. The gap between them was eroded by a lateral glacial stream when ice occupied the Spray River valley. Upon retreat of the ice from the Spray River valley, the Bow River was diverted through the gap, probably due to a glacial sediment dam between Tunnel and Stoney Squaw mountains. In the distance is the Cascade Coal Basin flanked on the east by the Fairholme Range rising above Lake Minnewanka.

Beautiful Lake O'Hara lies at the foot of Mount Victoria (left) and Mount Huber, each carved from Cambrian strata of the Gog Group and Mount Whyte, Cathedral, Stephen and Eldon formations. Above the far end of the lake is the hanging valley containing Lake Oesa.

The turquoise-blue waters of Lake McArthur result from low concentrations of very fine glacial flour suspended in the deep water.

From a viewpoint on the Icefields Parkway you get a glimpse of the Waputik Icefield which spawns Hector Lake, surrounded by gently inclined Cambrian strata of the Gog Group and Mt. Whyte, Cathedral, Stephen and Eldon formations. Beneath the lake, glacial varves record the annual accumulation of sediments beginning as far back as the dawn of civilization.

As you approach the "Big Hill" leading up to Sunwapta Pass and the Columbia Icefield you see ahead two mountains which, between them, reveal a synclinal fold. To the left, and closest to you, carbonate strata of the Palliser Formation form a gentle downward arc. The continuation of the syncline is seen on Nigel Peak in the distance where the Banff and Livingstone formations outline the remainder syncline.

A classic thrust sheet is shown my Mt. Rundle which displays the characteristic three-fold formation signature of the Front Ranges: Palliser Formation (lower cliff), Banff Formation (middle slope), Livingstone Formation (upper cliff). The surface trace of the Rundle Thrust occurs beneath the lower slopes. To the right the Spray River Valley is a classic U-shaped glacial valley.

Above Emerald Lake, the upper cliffs of Wapta Mountain are formed dominantly of carbonates of the Cambrian Eldon Formation, which occur above limy shales of the Stephen Formation forming Fossil Ridge to the right. Not distinguishable on the upper part of the ridge is Charles Walcott's famous quarry in the Burgess Shale where many of the world's most important fossils occur. (Photo by Lee McKenzie McAnally.)

These towers on the Wiwaxy Peaks developed between vertical fractures in Cambrian quartzites of the Gog Group and carbonates of the Cathedral Formation.

The prominent vertical wall of Mount Yamnuska is formed from Cambrian carbonates of the Eldon Formation. These have been thrust along the McConnell Fault over Cretaceous clastics of the Belly River Formation forming the lower slopes. This mountain is the easternmost of the front ranges which here are separated from the Foothills by the McConnell Thrust.

From the viewpoint above Peyto Lake, the valley of the Mistaya River stretches northward towards Mount Murchison in the distance. The river valley is formed along the surface trace of the Simpson Pass Thrust which approximates the course of the Icefields Parkway as far as Jasper. The peaks to the left are Mistaya Mountain and Breaker Mountain, formed mainly from Cambrian carbonate strata.

Viewed from the "Big Hill" leading up to Sunwapta Pass and the Columbia Icefield, the right, or west half of Cirrus Mountain exposes the classic, three-fold formational structure of the front ranges: lower cliffs of Palliser carbonate, middle slopes of Banff limy shale and upper cliffs of Livingstone carbonate. In the centre of the photograph a steeply inclined normal fault passes through the skyline notch.

Viewed from the Trans-Canada Highway near its junction with the Kananaskis Highway, the mountain front appears as a wall of carbonate rising above the subdued topography of the Foothills. To the far right is Mount Yamnuska, its steep cliff formed from Cambrian carbonates of the Eldon Formation, which have been thrust upward and over the soft shales of the Cretaceous Belly River Formation supporting the lower slopes. To the left (south) of Mt. Yamnuska the Eldon Formation is successively overlain by the Pika Formation and Devonian strata of the Cairn and Southesk formations, parts of the Fairholme Group. A thin yellow-weathering zone occurs just above the prominent break in slope about three-quarters of the way up the mountain and represents the unconformity between Cambrian and Devonian strata in this area.

Ranges of the Rockies. According to Ray Price, the rocks in the hanging wall above the fault have been shoved some 32 kilometres northeastward over the rocks in the footwall beneath the fault. A climb to the base of the cliff will bring you right to the fault where you can actually see the fracture between the two formations and stick your hand in it. From Highway 1A, close to the base of the mountain, if the light is right, you can see many inclined fine striations in the face of the cliff. These are tiny fractures, each one a minor thrust fault, that developed in the Eldon Formation carbonates during the period of thrusting. Surprisingly, the soft, underlying shales and siltstones of the Belly River Formation remained undisturbed.

On the rock-face to the left, or southwest of Mount Yamnuska, the McConnell Thrust occurs at the base of the massive-looking carbonates of the Eldon Formation, which is surmounted by the Cambrian Pika Formation. Above the Pika the thin yellowish layer about two-thirds of the way up the side of the mountain represents the uncon-

Across from the viewpoint on the Trans-Canada Highway at Lac des Arcs east of Canmore, the westerly inclined Lac des Arcs Thrust juxtaposes Cambrian strata of the Eldon Formation (left) above Carboniferous rocks of the Banff Formation (right). The surface trace of the thrust occurs just to the left of the small knoll in the centre of the photograph. Further to the right lies the La farge Company cement plant at Exshaw which quarries limestone from the Devonian Palliser Formation.

formity between Cambrian strata and the overlying carbonates of the Devonian Fairholme Group and Palliser Formation (*see page 22*).

LAC DES ARCS VIEWPOINT

From the viewpoint on the south side of Lac des Arcs you can look northward across the lake to the broad, low mountain lying to the west (left) of the Lafarge Company cement plant at Exshaw. The plant has been in continuous operation since 1906 and quarries pure limestone from the Devonian Palliser Formation which occurs above the Exshaw Thrust.

The low mountain to the west of the town displays westerly-inclined Paleozoic strata, which occur in thrust sheets of the Exshaw and Lac des Arcs thrust faults. The surface trace of the Lac des Arcs Thrust is on the left (west) slope of the low ridge in the centre of the photograph. Beneath the thrust, in its footwall, are Carboniferous strata of the Livingstone and Banff formations. Above the thrust, the hangingwall consists of dominantly carbonate strata of Cambrian,

Devonian and Carboniferous ages forming the Lac des Arcs thrust sheet.

Westward from Lac des Arcs you pass by several outcrops of Devonian carbonate strata that were constructed by reef-building organisms such as stromatoporoids and corals. As you turn northwesterly at Gap Lake between Grotto Mountain to the north and Pigeon Mountain to the south, you enter the Cascade Coal Basin containing coal-bearing strata in the Cretaceous Kootenay Group (*see page 77*). The surface trace of the Rundle Thrust occurs beneath the base of the cliff, which consists of folded Palliser Formation carbonate in the easternmost of the Three Sisters peaks to the southwest.

GRASSI LAKES PARK

To get to Grassi Lakes Park you drive southwestward through the town of Canmore, turn left (south) onto 8th Avenue, cross the Bow River and railway tracks and follow the Spray Lakes Road to where a sign points to an intersecting gravel road leading to the parking area. From there you walk up the well-marked trail for about twenty to twenty-five minutes until you come to the lovely green and cobalt-blue ponds nestling at the foot of a prominent mass of limestone. This is a remarkable place indeed, named in honour of Lawrence Grassi, a dedicated trail builder in the Rockies and, for many years, the much-loved park warden at Lake O'Hara.

What you are looking at across the ponds is an ancient reef. The reef is formed from the shells of lime-secreting organisms such as stromatoporoids, which, some 370 million years ago, built this structure in the warm, shallow waters covering the eastern miogeocline and adjacent platform. It closely resembles the tropical coral reefs of today. This reef, which developed in the Cairn Formation of the Fairholme Group, is identical to much larger structures that act as petroleum reservoirs beneath the Alberta plains. Oil fields such as Leduc, Redwater, Swan Hills, and many others produce petroleum from such reefs. If you walk along the path at the base of the large mound you will notice that the rock is full of holes. The holes resulted from the chemical change from calcite to dolomite in the centres of stromatoporoids, shortly after the reef was buried and killed.

This reef, like those beneath the plains, was buried over a period of many millions of years. The initial deposits that accumulated beside and over these reefs were black, limy shales containing abundant organic matter derived from decomposed cellular material of marine organisms. With progressively deeper burial the added pressure and

The massive carbonate reef at lovely Grassi Lakes Park above the town of Canmore is formed from shells of Devonian stromatoporoids and other fossils (top). The sponge-like character of the reef is evidenced in the many holes, or pores, which form an interconnected network of chambers and passageways (bottom). It is this type of structure which forms oil and gas reservoirs in identical, but much larger reefs beneath the Alberta plains.

Driving westwards from Canmore you pass beneath the rugged northeast face of Mount Rundle, formed of Devonian and Carboniferous strata within the Rundle Thrust Sheet.

heat caused the organic material to change into liquid and gaseous petroleum which later was squeezed out of the limy shales, ultimately becoming trapped in the sponge-like porous reef. In the case of the Grassi Lakes reef, the forces accompanying mountain building elevated this part of the miogeocline; consequently all of the surrounding and younger black shales that once covered the reef have been eroded away. If this reef ever contained oil and gas, they have long since disappeared.

From the trail at the base of the cliff you can get a good look at the several parts of the reef because, during uplift of the mountains, it was conveniently turned on its side. At its base, closest to the lake shore, the rocks consist of sediments which accumulated in shallow lagoons. These in turn are overlain by carbonate strata formed from the shells of stromatoporoids and corals that built the reef. You will have no difficulty in recognizing the organic origin of the reef as the strata at some places are composed entirely of fossils.

CASCADE COAL BASIN, HOODOOS AND CASCADE MOUNTAIN

The Cascade Coal Basin was the site of semi-continuous coal mining from the early 1880s until 1979 when the last mine at Canmore closed. Other small mining towns in the basin included Anthracite east of

From a viewpoint on Tunnel Mountain Road, glacial sediments can be seen to have been eroded by heavy rains into ghostly shapes resembling the silent stone faces of Easter Island.

Banff and Bankhead on the road leading to Lake Minnewanka; these mines closed operations in 1897 and 1922, respectively. The coals, which occur in the Lower Cretaceous Mist Mountain Formation of the Kootenay Group, are mainly of low-ash, low-sulphur, thermal type, most useful for thermo-electric power generation. Don Norris, formerly of the Geological Survey of Canada, carried out detailed studies of the basin and, in 1971, estimated that less than 200 million tonnes of recoverable reserves remained to be extracted.

Driving northwestward from Canmore along the axis of the Cascade Coal Basin the massive face of Mount Rundle is to your left (southwest). Near the bottom of the forested slopes at the base of the mountain is the surface trace of the Rundle Thrust. The uppermost cliffs are formed of Carboniferous Livingstone Formation carbonate, beneath which is the gentler slope of Banff Formation shales, then the steep cliffs of Palliser Formation carbonate. Strata of the Fairholme Group form the lowest slopes above the thrust fault.

Along both sides of the highway you will notice a series of pale grey bluffs of gravel and sand that have been sculpted into strange, ghostly-looking shapes, some reminiscent of the great stone faces of Easter Island. These are called **hoodoos**, formed from glacial sediments

Close to the junction of the Trans-Canada Highway and the eastern entrance to Banff townsite, steeply inclined ribs of sandstone and shale of the Upper Jurassic Fernie Formation represent the first sediments derived from erosion of the uplifted collision suture between North America and the Intermontane Superterrane (*see page 33*). These sediments accumulated in the trough that developed in front of the easterly advancing thrust sheets and were subsequently folded beneath the Rundle Thrust to form the overturned southwest limb of the Mt. Allan Syncline. Cascade Mountain forms the background. (Photo by W.R. Price.)

and carved by the erosive action of rain and storm run-off during the past several thousand years. The generally accepted view is that the glacial sediments consist of outwash gravels deposited by a braided river flowing from a valley glacier and an overlying succession of till deposited beneath, and moulded by, an advancing glacier during the Bow Valley Advance (*see page 42*). An alternative, but not widely accepted hypothesis, is that the sediments from which the hoodoos have been carved are the products of debris flows. In this model it is argued that between 10,000 and 12,000 years ago, as the Bow Valley glacier receded, the waters of ice-dammed lakes in tributary valleys were suddenly released resulting in the remobilization of large volumes of former glacial outwash. This material was swept downstream along the Bow Valley to be redeposited as a thick layer of braided river sediments from which the hoodoos have been sculpted. Ahead of you is the massive southeast face of Cascade Mountain which exposes the Palliser, Banff and Livingstone formations above the surface trace of the Rundle Thrust (*see photo on page 65*).

On approaching Cascade Mountain you will see some steeply inclined strata along the roadside to your right. These are clastic rocks of the Upper Jurassic Fernie Formation and Lower Cretaceous Kootenay Group which occur within the Cascade Coal Basin. These strata are situated beneath the Rundle Thrust where they form one side, or **limb**, of the Mount Allen Syncline. When the Paleozoic rocks forming Mount Rundle and Cascade Mountain were shoved upwards and overtop of the Fernie and Kootenay strata, the latter were dragged and bent upwards past the vertical such that they are now nearly upside-down. The upper part of the Fernie Formation as well as the overlying Kootenay Group are composed of clastic sediments that were eroded from sources far to the west, including the Omineca Belt and the earliest, westernmost thrust sheets of the Rockies.

LAKE MINNEWANKA

The name "Minnewanka," an Assiniboine word meaning "lake of the water spirit," was first used by the Department of the Interior in 1888 to replace the existing names "Devil's Head Lake" and "Cannibal Lake." The latter of these earlier names relates to an Indian legend concerning a cannibal ghost that once was observed picking over the bones of Stoney (Assiniboine) warriors killed in battle on the nearby Ghost River. Expansion of the lake was accomplished by construction of the Devil's Canyon Dam which, in 1912, formed a reservoir some five metres deep. In 1941 an additional twenty metres of storage was added thus creating the modern lake which today is eleven kilometres long, the largest in Banff National Park, and which, at its eastern end, adjoins the Ghost Lakes (not to be confused with Ghost Lake, a reservoir on the Bow River). During the late stages of Wisconsinan glaciation when the Bow Valley was filled with ice, the main drainage of this area was through the valley occupied by the lake (*see page 43*).

The imposing mountain on the southeast side of the lake is Mount Inglismaldie which, together with Mount Girouard, Mount Peechee, Princess Margaret Mountain and Grotto Mountain, form the Fairholme Range. Mount Inglismaldie is sculpted largely from the same three formations that constitute Mount Rundle and Cascade Mountain, namely the lower, cliff-forming Palliser Formation of Devonian age, the middle recessive slope-forming Banff Formation of Carboniferous age and the upper cliffy Livingstone Formation of the Rundle Group, also of Carboniferous age.

Above the south shore of Lake Minnewanka, Mount Inglismaldie is made up of carbonates of the Palliser Formation forming the lower cliff, shaly limestones of the Banff Formation underlying the middle slopes and carbonates of the Livingstone Formation at the top. The mountain is named after Inglismaldie Castle in Kincardineshire, Scotland, seat of the Earl of Kintore whose visit to Banff in 1886 inspired the park superintendent to name the mountain in his honour.

BANFF TOWNSITE, BOW FALLS AND THE SULPHUR MOUNTAIN GONDOLA

A decade after the first visit by Europeans to what became the Banff Hot Springs, the townsite of Banff was founded in 1886, one year after the establishment of Banff National Park. The town is situated at the confluence of the Bow and Spray rivers and is built upon river alluvium and glacial deposits. Between Tunnel Mountain and Mount Rundle the Bow River at Bow Falls cascades over Triassic strata of the Sulphur Mountain Formation, which has supplied much of the distinctive stone for building facades of the town, including those of the baronial Banff Springs Hotel. Over the past many decades Banff has been well known for its two hot springs, the Cave and Basin and the Upper Hot Springs, both of which occur close to the surface trace of the Sulphur Mountain Thrust. Due to the **geothermal gradient**, temperature increases with depth by approximately 25°C per kilometre; thus surface waters, which percolate downwards to depths of between three and four kilometres, are heated to high temperatures. At these depths the hot water dissolves compounds of calcium, magnesium

Below the Banff Springs Hotel the Bow River cascades over steeply inclined Triassic strata of the Sulphur Mountain Formation.

and sulphur before returning to the surface along the Sulphur Mountain Thrust, its rate of flow aided by cavities dissolved out of carbonates of the Devonian Cairn Formation along the hanging wall of the thrust fracture. The average temperature of the Upper Hot Springs is about 38 degrees C, whereas that of the Cave and Basin is somewhat cooler at 30 degrees C. An undeveloped hot spring occurs along the thrust about an hour's walk westward from the park administration building. There you will find a shallow cave within a deposit of **tufa**, a spongy mass of porous and soft calcium carbonate and gypsum; tufa is formed when hot water that is saturated with these compounds cools, thus precipitating them out of solution. Several small pools of warm water occur in the vicinity, as does the distinctive "rotten-egg" smell of hydrogen sulphide.

The Sulphur Mountain Gondola, next to the Upper Hot Springs, takes you on an eight-minute ride, beginning at the surface trace of the Sulphur Mountain Thrust, up some 700 metres of vertical relief, crossing the Devonian Cairn, Southesk, Alexo and Palliser formations to the upper terminal constructed upon shaly carbonates and shale of the Carboniferous Banff Formation (*see photo page 67*). The views from the upper terminal and summit trail are spectacular. Directly below, you can see how erosion has separated the Rundle Range into two

As viewed from the upper terminal of the Sulphur Mountain Gondola, the south face of Mount Norquay and nearby Mount Edith display the classic style of successive, shingled thrust sheets. To the right the east-facing slope of Mount Norquay exposes Paleozoic carbonate and shale strata of the Fairholme Group, Palliser and Banff Formations, and Rundle and Rocky Mountain groups in the hanging wall of the Sulphur Mountain Thrust, the surface trace of which lies just below the upper terminal of the Norquay chairlift. Beneath the thrust are Triassic strata of the Spray River Group. On the east slope of the ridge of Mount Edith, to the left, the Bourgeau Thrust has juxtaposed the same stratigraphy: Devonian and younger strata above the thrust have been shoved overtop of Triassic strata forming the recessive slope at the bottom of the valley between the two peaks.

chunks, Tunnel Mountain to the northwest and Mount Rundle to the southeast. The erosion is thought to have been due to a stream flowing along the contact between the flank of Mount Rundle and an ice tongue occupying Spray River Valley. As the ice melted the stream cut its way through the Rundle Range. Later, as a consequence of ice remaining between Tunnel and Stoney Squaw mountains or glacial outwash blocking its course, the Bow River was diverted through the gap (*see page 43*). In the distance to the east the Fairholme Range forms the east side of the Cascade Coal Basin (*see page 77*).

To the northwest is the rugged, serrated ridge of the Sawback Range. The narrow valley between it and Mount Norquay is formed along the surface trace of the Borgeau Thrust, above which Devonian strata of the Borgeau Thrust Sheet have been overridden by another thrust sheet composed of Cambrian to Carboniferous rocks beneath the Sawback Thrust. The steeply dipping strata of the Sawback Thrust Sheet form the main bulk of the range which, on its western flank, is breached by

Hole-In-The-Wall, a cave dissolved in the limestone some 480 metres above the floor of the Bow Valley (*see page 45*).

MOUNT RUNDLE

The best place to view the grandeur of Mount Rundle is from a lovely roadside meadow on the southeast slope of Mount Norquay. As you drive up the Norquay Road from its intersection with the Trans-Canada Highway, outcrops along the way are formed from clastic strata belonging to the Triassic Spray River Group.

Mount Rundle gets its name from Robert Terrill Rundle, an itinerant Wesleyan missionary whose exploits and travels throughout western Canada during the middle of the nineteenth century, like so many others of his day, read like a study in flies, ghastly weather and exhaustion. It seems to me that there are other people at least as deserving, after whom the mountain could have been named, but I like the way "Rundle" tumbles off the tongue. The name seems to go with the mountain (*see photo on page 70*).

Looking across the Bow River valley, Mount Rundle appears like a massive guardian fortress above the town of Banff. Perhaps Earl Birney had Mount Rundle in mind when, for his tragic poem *David*, he wrote:

> "... *The peak was upthrust*
> *Like a fist in a frozen ocean of rock that swirled*
> *Into valleys the moon could be rolled in.*"

Mount Rundle is a textbook example of the geological architecture of the Front Ranges of the Rockies. The rocks from which it has been carved are characteristic of the kinds of sediments laid down upon the floor of the shallow sea that covered the miogeocline for over 400 million years, and the form or shape of the mountain represents the quintessence of thrust sheets.

The lowermost cliff of Palliser Formation carbonate preserves a record similar to the environment currently found behind reefs of the modern Bahama Banks in the Caribbean, where waves and currents transport fine lime muds in shallow water. Between the Palliser and the steep slopes of the Banff Formation is the Exshaw Formation, a thin succession of black, organic shale which occurs widely throughout North America (*see page 26*).

The Banff Formation, of Early Carboniferous age, consists of limy shales and shaly limestones, locally as much as 1,200 metres thick.

Above the Banff are the imposing carbonate cliffs of the Livingstone Formation of the Rundle Group. The Livingstone consists largely of broken shell material which is thought to have accumulated as broad shell banks on the eastern miogeocline and platform (*see page 27*).

The form of Mount Rundle, with its steep eastern face, tilted strata and gently inclined western slope, is that of a classic thrust sheet. This sheet, which also includes Cascade Mountain and the Three Sisters, was shoved eastward along the Rundle Thrust, which occurs beneath Fairholme Group and Cambrian strata in the the lower slopes of the mountain. In the footwall, beneath the fault, lie the clastic sediments of the Cascade Coal Basin you saw earlier (*see page 79*). The sweeping back slope, forming the valley "the moon could be rolled in" is closely parallel to the next thrust fault to the west; above lie the strata of Sulphur Mountain which repeats the same three formations: Palliser, Banff and Livingstone (*see page 52*).

SAWBACK RANGE TO CASTLE MOUNTAIN VIEWPOINT

Returning to the Trans-Canada Highway you proceed southwestward then northwestward around the southeastern snout of Sawback Range. Earlier in the book (*see page 52*) I spoke of what could be seen from the Vermilion Lakes Viewpoint, which regrettably is inaccessible to those travelling westward. The Sawback Range hosts a typical and thick succession of Paleozoic strata which occur above the Bourgeau Thrust and are dislocated by several normal faults.

Rounding the south end of Sawback Range you can see the cave known as Hole-In-The-Wall, developed in the cliff of Livingstone Formation carbonate on the southward projecting spur of Mount Cory (*see page 45*). From here to the viewpoint opposite the south end of Castle Mountain, the course of the highway is approximately parallel to the trend of Sawback Range. To your right the southwest flank of the range appears as a nearly vertical slab of Paleozoic carbonate rocks.

CASTLE MOUNTAIN VIEWPOINT

This viewpoint beside the Bow River provides an outstanding view of the southern buttresses of Castle Mountain, the easternmost of the Main Ranges at this latitude.

The mountain's castellated succession of Middle Cambrian strata lies above the Castle Mountain Thrust, the surface trace of which occurs in the forested slopes beneath the lower cliffs. This thrust, a splay off the Simpson Pass Thrust, here serves as the boundary between the

Clouds enshroud the Cambrian carbonate turrets forming Eisenhower Peak on Castle Mountain.
Photo by K. McCormick.

Front and Main ranges. The tree-covered slopes are formed from quartzites of the Gog Group. The lower cliffs consist of a thin succession of clastic strata belonging to the Mount Whyte Formation which, in turn, are overlain by thick carbonates of the Cathedral Formation. A distinctive, thin shaly carbonate called the Stephen Formation separates the lower from the upper carbonate cliffs, the latter consisting of the Eldon and Pika formations (*see cover photo*).

Jim Aitken, introduced earlier as the author of the Kicking Horse Rim idea (*see page 18*), recognized that the Cambrian rocks of the Rockies repeated their overall compositions through time in a series of what he called "grand cycles." Three such cycles are displayed in Castle Mountain, with the Mount Whyte (clastics) and Cathedral (carbonate) formations comprising the lowest cycle, the Stephen (clastics) and Eldon (carbonates) formations the middle cycle and the Pika Formation (lower clastics and upper carbonates) the top cycle. Each cycle begins with clastic rocks which are then succeeded by carbonates. The clastics are the product of erosion of the interior Precambrian core of the continent, specifically the Canadian Shield. These appear to have been deposited in deeper, offshore waters as opposed to the carbonates which accumulated in shallower waters closer to the shoreline. As the sea repeatedly advanced and retreated across the miogeocline these different sediment types were deposited in adjacent environments such that, at any one place, the succession of clastic to carbonate strata reflects the cyclical and repetitive nature of the processes leading to their accumulation.

These cycles serve to illustrate the repetitive nature of many geological processes such as glaciation, the formation and subsequent break-up of supercontinents, catastrophic faunal extinctions and many others. Several years ago Dr. Peter Gretener of the University of Calgary published an interesting paper called "Significance of the Rare Event in Geology" (*see Sources and Additional Reading*) in which he argued that given enough time, the improbable becomes probable and eventually approaches certainty. If a natural event is known to have happened once, then most likely it has happened at least twice and/or will happen again.

Behind, or on the east flank of Castle Mountain, is a large cirque within which is Rockbound Lake, a tarn (*see page 41*). This amphitheatre-like depression, like many others throughout the Rockies, was carved by alpine glaciers during and since late Wisconsinan glaciation.

THE MAIN RANGES: CASTLE MOUNTAIN TO LAKE LOUISE

Driving northwesterly towards Castle Mountain you will see evidence of landslides. For example, close to the southeast end of the mountain the hummocky topography along the Bow River represents a landslide of Triassic siltstone that cascaded into the valley from the scar which is visible on the west slope of Sawback Range. Farther to the northwest, on the southwest slope of Castle Mountain, are large slide blocks which collapsed from carbonate cliffs of Eldon and Cathedral carbonate that form the top of the mountain and which came to rest on the lower slopes of Miette Group strata.

At the junction with Highway 93 you can turn southeasterly onto Highway 1A and drive about six kilometres to Johnson Canyon. A walk along the narrow gorge being carved through carbonate of the Carboniferous Livingstone Formation takes you past two impressive waterfalls. Several interpretive signs outline the developmental history of the canyon, beginning with a landslide off Mount Ishbel, that is estimated to have occurred some 8,000 years ago (*see page 43*).

Along the stretch of the Trans-Canada Highway, from its junction with Highway 93 to Lake Louise, you travel parallel to the course of the surface trace of the Simpson Pass Thrust. This thrust fault is one of the longest in the Rockies, extending from well south of Banff to north of Jasper, a total distance of over 300 kilometres. This fault is

On the southwest slope of Castle Mountain, huge slide-blocks of Cambrian Eldon and Cathedral carbonates have collapsed upon Precambrian slates of the Miette Group.

discussed further in the section on the Icefields Parkway from Lake Louise to the Columbia Icefields.

As you approach Lake Louise the prominent mountain on your left is Mount Temple. The lower cliffs are composed of Lower Cambrian clastics of the Gog Group, the upper cliffs comprise Middle Cambrian carbonates of the Cathedral, Stephen and Eldon formations. Roadside outcrops consist of slates belonging to the Precambrian Miette Group. It is these sediments which accumulated as a thick blanket of strata across the rifted edge of the continent when North America separated from Australia some 750 million years ago (*see page 15*).

LAKE LOUISE AND MORAINE LAKE

It is hard to imagine any other place on earth where two such incredibly beautiful settings can be found. Indeed, this general region, embracing Lake Louise and Moraine Lake and including Lake O'Hara and Lake McArthur in Yoho National Park (*see page 92*), is probably unsurpassed, for spectacular scenery, by any other alpine region in the world.

Lake Louise was named in honour of Her Royal Highness Princess Louise Caroline Alberta (1848 – 1939), fourth daughter of Queen Victoria and wife of the Marquis of Lorne, later 9th Duke of Argyle, who was Governor General of Canada from 1878 to 1883 and founder of the Royal Society of Canada and the National Gallery (*see photo on page 66*). On the locality map on page 5, Lake Louise is situated between Mount Niblock (21) on the Continental Divide and the village of Lake Louise.

Lake Louise is possibly a tarn, its basin carved by the Victoria Glacier at a time when the glacier was much larger and extended northeastwards down the valley of the Plain of Six Glaciers, perhaps during the Bow Valley Advance, some twenty-five thousand years ago (*see page 43*), and again later during the "**Little Ice Age**" between about 1500 AD and the mid-1800s (*see page 98, 120 and Glossary*). The Chateau Lake Louise, at the lake's northeastern end, is situated on a **recessional moraine** that was deposited upon a rock sill. Like other lakes in the Rockies its fabulous colour is due to a combination of dissolved compounds and, more important, to the presence of extremely fine rock flour particles suspended in the water (*see page 3*).

Evidence of the glacial history of the lake basin is provided by abundant glacial **striations** etched into scoured bedrock and by narrow, sharp-crested lateral moraines along the margins of the somewhat desolate Plain of Six Glaciers. From there one gets spectacular views of Mount Victoria and Mount Lefroy, the crest of each carved from thick carbonates of the Cambrian Eldon Formation. Separating these two mountains is Abbot Pass, formed along the course of a normal fault which extends southward to intersect the Cataract Brook Fault, another normal fault that traverses Lake O'Hara valley (*see page 96*). Glaciers visible from the Plain of Six Glaciers, are from right to left: Pope's Glacier, Victoria, Lefroy and Aberdeen Glacier. As noted earlier, these glaciers are the products of snow accumulation on the northeast-facing slopes of the ranges, in this case the Bow Range.

The rocks enclosing Lake Louise basin are Precambrian and Cambrian in age, consisting of slate and gritty sandstone of the Miette Group and quartzite of the Gog Group, the latter seen on The Beehive west of the hotel and on Fairview Mountain to the south. On distant peaks surrounding Victoria Glacier the Gog is overlain by the Mount Whyte, Cathedral, Stephen, Eldon and Pika formations.

Moraine Lake, in the Valley of the Ten Peaks, south of Lake Louise, is enclosed within a broad valley between Mount Temple and Pinnacle Mountain on the north and the row of ten peaks to the south and southwest. These magnificent peaks, all of which are carved from horizontal Cambrian strata of the Gog Group and the Cathedral, Stephen and Eldon formations, are, from east to west: Mount Fay and

On a clear winter day Moraine Lake lies frozen beneath Cambrian quartzite strata of the Gog Group and dominantly carbonate strata of the Cathedral, Stephen and Eldon formations which together make up the surrounding mountains of the Valley of the Ten Peaks. (Photo by K. McCormick.)

The emerald-green waters of Moraine Lake are dammed by a landslide of Gog Group quartzite that may have fallen from Mount Fay and the Tower of Babel, the lower cliffs of which are seen at the right. (Photo by Ben Gadd.)

the Tower of Babel, Mount Little, Mount Bowlen, Mount Perren, Mount Septa, Mount Allen, Mount Tuzo, Deltaform Mountain, and Neptuak and Wenkchema peaks.

In terms of its origins, the name "Moraine" Lake is partly misleading. The lake is dammed by a small landslide-pile of Gog quartzite which can be clearly seen at its northeast end. The source of the landslide might have been the steep slopes of adjacent Mount Fay and the Tower of Babel or from farther up the valley. If the latter was the case, then a glacier extending down the valley may have pushed or carried the landslide debris to its present position, thus making it a moraine.

LAKE LOUISE TO THE CONTINENTAL DIVIDE

From the village of Lake Louise you see the Whitehorn ski development on the southwest-facing slope of the Slate Range, carved from Miette Group strata. At the junction of the Trans-Canada Highway with the Icefields Parkway (Hwy 93) are gently warped slates belonging to the Corral Creek Formation of the Miette Group. The Corral Creek Formation, in addition to slates, consists of a variety of rocks including sandstone and conglomerate and contains the oldest strata known in the two parks.

THE CONTINENTAL DIVIDE

The boundary between Alberta and British Columbia (the boundary between Banff and Yoho national parks) coincides with The Continental Divide separating two of the three watersheds of the North American continent (*see page 2*). Approaching the divide you may have noticed that the waters of Bath Creek beside the highway were flowing eastward to join the Bow. Continuing on westward, however, you will soon see that the Kicking Horse River drains westward from Wapta Lake, ultimately to join the Columbia River at Golden. Here, the area embracing the Continental Divide occurs within Kicking Horse Pass between Mount Niblock on the south and Mount Bosworth to the north. You can look up at the peaks to the south and follow the course of the Continental Divide along the ridge between Mount Niblock on the left and Narao Peak on your right. From the ridge the divide descends toward you along the rounded shoulder of Naro Peak to cross the highway.

From the Continental Divide you pass by Wapta Lake, the headwaters of the Kicking Horse River. The lake itself is fed by Cataract Brook which drains northward from Lake O'Hara. At the western end of

Wapta Lake you cross the surface trace of the Cataract Brook Fault, a normal fault across which the west side drops downwards some 1,400 metres relative to the rocks on the east side. To the north of the lake the fault traverses the east side of Paget Peak which exposes strata of the Pika, Arctomys, Waterfowl, Sullivan and Lyell formations, all of Middle and Late Cambrian age. Along the highway as far as the Spiral Tunnels Viewpoint, Lower Cambrian strata belong to the Eldon, Stephen, Cathedral and Mount Whyte formations and the Gog Group.

Lake O'Hara and Vicinity

A widely held belief, and one for which I have much sympathy, is that the region surrounding and including Lake O'Hara is the most scenically magnificent area in the Canadian Rockies. On the somewhat crowded location map shown on page 5, the Lake O'Hara area lies west of the Continental Divide in Yoho National Park, south of Cathedral Mountain (8) and east of Mount Dennis (13). Following the short eleven-kilometre bus ride from the parking lot just east of Wapta Lake (junction of Highways 1 and 1A) the visitor can stay either at a comfortable lakeside lodge, or in one of several small lakeside cabins, or else "tough it out" in the campground where prepared pads for thirty tents are available through reservations made with the Yoho park service (250-343-6433). From these central locations the visitor can hike twenty-nine different and excellently maintained trails requiring variable lung capacity, leg muscles and determination. At the low end of the scale is the easy and pleasant walk around Lake O'Hara and, at the upper end, the vigorous scramble up to Abbot Pass overlooking the Victoria and Lefroy glaciers above Lake Louise.

The following account of the Lake O'Hara area derives from a camping holiday with my wife, Linda, in late August of 1995. Despite having visited the lake some forty-five years earlier, I had little recollection of what we were to see. Our weather was variable: one magnificently sunny day followed by two days of intermittent rain and sunshine. At night it poured. As always when backpacking in the Rockies, one of the most interesting things to observe is what other people eat. In the evenings we would crowd into one of the two covered shelters to cook our supper of dried something-or-other over a small primus stove (this after a happy hour of Tang mixed with overproof rum — a taste sensation not recommended). The array of cooking gear and food varieties was astonishing. One young couple fried corn-on-the-cob while another prepared a sumptuous feast of pasta, Westphalia ham and asparagus sauce. The campground was truly cosmopolitan with the

Looming through the mist, like some forbidding Draculean castle, the turreted edifice of Mount Huber has been moulded from Cambrian strata of the Cathedral, Stephen and Eldon formations.

sounds of French, German, Japanese and English mixing in a delightful cacophony of spirited conversations about the adventures of that day and those expected on the next. It's a fun place.

The spectacular scenery of the Lake O'Hara region is partly a consequence of the manner by which the rocks forming the mountains responded to the processes of glaciation and post-glacial erosion (*see photo on page 68*). The horizontal Cambrian strata forming the mountains, from bottom to top, consist of thick beds of quartzite of the Gog Group underlying the lower slopes and forming the basal cliffs; the thin Mount Whyte Formation shales and siltstones; and the Cathedral, Stephen and Eldon formations which make up the main upper buttresses. It is the Cathedral Formation which gives the region much of its unique character. The Cathedral consists of prominently layered limestone and dolomite which is vertically jointed, or fractured, such that it tends to form isolated turrets or battlements like those atop the Gog Formation on Wiwaxy Peaks (*see photo on page 71*). On Mount Huber the effect has been to create a series of tiered cupolas which, when enveloped in mist and low cloud, appear as gothic battlements of a Draculean castle looming above the O'Hara valley.

The many trails in the area provide a glimpse of the Cambrian world, as it was some 550 million years ago. By closely observing the

1

2

3

4

By looking closely at the surfaces of slabs of limestone in the Lake O'Hara area you can get some idea of the kind of world that existed during the Cambrian Period, some 550 million years ago. Feeding trails (1), stromatolites (2), sun cracks (3) and ripple marks (4) suggest a shallow, clear, warm sea, covering a muddy bottom upon which many creatures foraged for food. Occasionally when the sea floor was exposed to the air, it dried in the sun.

slabs of talus across which you are walking – blocks which have fallen from the cliffs above – you can see many features indicative of the shallow sea covering the miogeocline at that time. For example, on the surfaces of many carbonate slabs you can see a wide variety of markings left by grazing and burrowing crustaceans and worms that foraged on the ancient sea bed. The burrows and trails were later filled with sediment, thus making them easily seen on the rocks. On some slabs you can see polygonal shapes formed when the shallow sea bed was briefly exposed to the sun, allowing the sediment to dry and shrink like mud in a drying mud-puddle. Many slabs show ripple marks, parallel ridges and valleys, formed as marine currents moved sediment across the sea bottom in the same way as does the rising and falling tide on a modern beach. Some ripple marks are of the interference type, with complex shapes resulting from two or more opposing currents interfering with one another. On the trail to Oesa Lake I saw **stromatolites** in a slab of Cathedral carbonate. Stromatolites are domal or columnar structures built by cyanobacteria in very shallow lagoons or tidal waters in tropical settings. The best known modern locality where such structures are growing is in Shark Bay in Western Australia where the very high salinity of the clear, warm water allows the lime-secreting bacteria to flourish in an environment free of bottom-scavenging organisms; such a setting was most probable during the early stages of Cathedral Formation accumulation. The picture that emerges from these observations is one of a shallow warm sea across which wind-driven currents moved sediment about on the sea bottom where different kinds of crustaceans and other creepy crawlies scavenged for food. Occasionally the sea was so shallow that the bottom periodically was exposed to the air at which times it dried to form sun cracks. In isolated clear pools, protected from the wind and drifting sediment, colonies of stromatolites built their structures in highly saline water, in exactly the same way they still do in Shark Bay, some 550 million years later.

The structural style of the Lake O'Hara area is dominated by normal faults. As described earlier (*see page 55 and Glossary*), normal faults are steeply-inclined fractures along which the block of crust on one side of the fault has dropped downward relative to the block on the other side. This means that in areas of stratified, sedimentary rocks such as the Rockies, normal faults juxtapose younger rocks in the down-dropped block against older rocks on the other side of the fault. As opposed to thrust faults which form as a consequence of compression, normal faults are a product of tension in the crust. Several normal faults disrupt the stratigraphic layering in the Main Ranges where Lake

The V-shaped valley of Cataract Brook is formed along the surface trace of the Cataract Brook normal fault. Rocks on the west (left) side of the fault have been dropped downwards some 600 metres with respect to the strata on the left (east).

O'Hara and Lake Louise are situated. The most prominent of these is the Cataract Brook Fault, the surface trace of which approximates the course of Cataract Brook which flows northward from Lake O'Hara into Wapta Lake. In the Lake O'Hara area the fault splays into two fractures, one on either side of Wiwaxy Peaks, which then rejoin to pass through the eastern end of the lake; from there it continues southeastwards across the Opabin Plateau and into Prospectors Valley in Kootenay National Park. Along this fault, near where it crosses Hungabee Mountain, Ray Price estimates that the rocks on the west side have been dropped downwards relative to those on the east side by about 600 metres, considerably less than his estimate of displacement near Wapta Lake (*see page 92*). Other normal faults include one extending southeasterly from Mount Schaffer through Lake McArthur and Biddle Pass, and two northerly tributaries to the Cataract Brook Fault one of which runs beneath Horseshoe Glacier on the east slopes of Hungabee Mountain and Ringrose Peak while the other passes through Yukness Mountain, beneath Oesa Lake and through Abbot Pass to beneath Victoria Glacier. All of these normal faults are believed to have formed in the latter stages of, or after, mountain-building, some 60 million years ago.

Another interesting aspect of the region is its glacial and post-gla-

Distant Lake O'Hara lies below a hanging valley enclosing a series of tarn lakes, landslides and moraines near Oesa Lake.

Oesa Lake is a tarn, occupying a basin quarried by glacial ice, beneath the snowbound ridge forming the Continental Divide.

cial history. On Opabin Plateau and around Oesa Lake a tangle of glacial moraines and landslides forms much of the topography within hanging valleys above Lake O'Hara. Oesa, O'Hara and McArthur lakes are tarns, occupying basins scooped out by the quarrying power of cirque-forming glaciers. A comparison of the colours of these three lakes at the time we were there illustrates the relationships between colour and suspended sediment-particle concentration and grain-size discussed earlier (*see page 3*). Oesa Lake and Lake O'Hara reflected a brilliant turquoise in the afternoon sun, although the colour of Oesa Lake was greyer due to a greater amount of coarser material in addition to rock flour particles, being closer to its glacial source than is Lake O'Hara. Lake McArthur, on the other hand, displayed a magnificent deep, cobalt blue (*see photo on page 68*). This is because the lake is larger and much deeper than the others at eighty-five metres (Lake O'Hara is forty-two metres deep), thus effecting a reduction in concentration and grain-size of fine suspended material in the uppermost part of the water column. Although most of the glacial topography was developed prior to about 10,000 years ago (*see page 41*), much of it probably reflects the Little Ice Age when, between about 1500 and 1850 AD, many valley glaciers in the Rockies advanced down their valleys, leaving lateral and terminal moraines in their wakes. In the Oesa Lake area, some evidence for recent glacial advance is shown by landslide debris which appears to have been pushed and moulded by moving ice. On the west side of Lake O'Hara a lateral moraine is estimated to be about 6500 years old but may be somewhat older. Lake O'Hara has a rock sill at its northern end which, together with landslide debris, helps to confine the lake.

One evening we went to a lecture on the post-glacial history of Lake O'Hara given by a geologist who did his master's thesis in the region. Mel Reasoner explained how, throughout the past 10,100 years, since the retreat of Wisconsinan ice of the Cordilleran Ice Sheet (*see page 42*), the lake has accumulated about 2.5 metres of organic sediment which accumulated on top of glacial material. In sediment cores he collected, two thin layers of light grey volcanic ash were enclosed by the dark organic muds. The upper layer is the Bridge River Ash which was ejected from the Meager Mountain volcano north of Vancouver some 2,350 years ago and carried by westerly winds as far as Alberta. The lower layer, identified as the Mazama Ash, originated from the eruption of Mount Mazama that created Crater Lake in Oregon some 6,800 years ago. Radiocarbon dates obtained from pine needles extracted from the cores immediately above the glacial sediments indicate that deglaciation of the region began about 10,100 years ago.

SPIRAL TUNNELS VIEWPOINT

The two spiral tunnels of the Canadian Pacific Railway were constructed to allow trains to climb from the deep valley of the Yoho River upwards to Kicking Horse Pass. The designer was John E. Schwitzer, a young engineer from McGill University who, knowing of their success in the high mountain passes of Europe, was the first to apply the concept in North America. Both tunnels were excavated through Lower Cambrian quartzite of the Gog Group, a tough form of sandstone forming the lower slopes of Cathedral Crags. The lower tunnel extends through an arc of 217 degrees and the upper through 291 degrees. In the process the rail bed is raised some thirty-two metres in a distance of less than two kilometres.

To the north, along the valley of the Yoho River, the surface trace of the Stephen-Cathedral Fault, another normal fault, traverses the western slope of the valley. It is estimated that the displacement along this fault is about 700 metres, whereby rocks of the Eldon Formation (525 million years old) slid downwards and came to rest beside rocks of the Gog Group (570 million years old).

MINE ADITS ON MOUNT FIELD

North of the intersection of the highway with the road to Takakkaw Falls you can see mine portals in the cliffs of Mount Whyte Formation shale and siltstone on the lower south slopes of Mount Field. The Kicking Horse Mine, together with the Monarch Mine on Mount Stephen on the south side of the highway and the nearby Silver Giant Mine, produced over 800,000 tonnes of lead/zinc/silver ore from the Cathedral Formation until their closure in 1952. These kinds of deposits are common features of facies changes from carbonate to shale, particularly in the northern Rockies and Mackenzie Mountains.

TAKAKKAW FALLS

The road to Takakkaw Falls, from where it intersects with the Trans-Canada Highway at the bottom of the big hill east of Field, takes you past the junction of the Yoho and Kicking Horse rivers, through a pair of nasty switchbacks and along the course of the Yoho River to the base of a cliff formed of prominently stratified limestone of the Upper Cambrian Lyell Formation. Originating as meltwater of the Daly Glacier not far beyond the top of the cliff (*D on location map, page 5*), the waters of Takakkaw Falls plunge in a narrow cataract some 380 metres into the valley of the Yoho River.

The cataract of Takakkaw Falls plunges 380 metres over a cliff formed from horizontal carbonate strata of the Upper Cambrian Lyell Formation. The source of the waterfall is the Daly Glacier, part of the Waputik Icefield on the Continental Divide.

FIELD LOOKOUT

A short distance west of the turn-off to the town of Field is a parking area where you can see the dramatic change in rock types and structural style from the eastern to the western Main Ranges of the Rockies. Looking to the southeast you see two mountains: Mount Stephen to the left and Mount Dennis to the right. The strata of both mountains are essentially the same age, Middle Cambrian. Those on Mount Stephen consist of thick-bedded, pure carbonates of the Eldon Formation whereas those on Mount Dennis, of the same age, are composed of thin-bedded, limy shale of the Chancellor Formation. The strata on Mount Dennis show abundant evidence of land-sliding and fragmentation, which probably occurred soon after they were deposited. If you hiked up the gully separating the two mountains into the alpine valley between Mount Stephen and Mount Duchesnay, you would actually be able to see where Eldon carbonate strata pass laterally into Chancellor shale. This change in rock type, or facies, between Mount

Mount Stephen and Mount Dennis lie on either side of the line of facies change in Cambrian strata from shallow water carbonate rocks on Mount Stephen to slumped, deeper water muddy limestone and shales on Mount Dennis. In Burgess Pass, between Mount Field and Wapta Mountain, this change in rock type coincides with the Cathedral Escarpment (Kicking Horse Rim) where, in the deep-water rocks, the famous soft-bodied fossils of the Burgess Shale are preserved. Drawing by Eric Yorath based upon a photo by W.R. Price.

Stephen and Mount Dennis represents the dramatic change in environment on the miogeocline, some 550 million years ago, from shallow, clear marine waters to the east to deeper, muddy waters to the west near the ancient edge of the continent. It is here where Jim Aitken first defined the so-called Kicking Horse Rim (*see page 18*).

To the northeast is Mount Field and, next to it to the southwest, Mount Burgess. Separating the two mountains is Burgess Pass. The famous fossil locality, the Burgess Shale, is situated on the western slope of Mount Field.

There are two aspects of the Burgess Shale which have been, and continue to be, subjects of intense research and debate. The first of these focuses on how the soft tissues of these 530-million-year-old creatures came to be preserved at all, the second on what they mean in terms of faunal evolution. With regard to the first of these, what is needed is a model to explain how a marine, bottom-living fauna could thrive in such numbers and diversity of form, and come to be preserved with its soft tissues intact. Under normal circumstances, when marine animals die the dissolved oxygen in the water combines with carbon in the soft tissues, resulting in their complete destruction. Moreover, bottom-scavenging organisms quickly consume such material as food, leaving only indigestible hard parts to be preserved as fossils. It seems therefore that the environment in which the Burgess animals lived had to have been different from that in which they died. Although not accepted by everyone, the most favoured hypothesis for the origin of the Burgess Shale and its fossils has to do with Jim Aitken's Kicking Horse Rim, which, in the vicinity of Field and Burgess Pass, is thought by Jim and his colleague Bill Fritz of the GSC to have comprised a steep submarine cliff, formed of the Cathedral Formation, and named by Bill the "Cathedral Escarpment." Jim has suggested that the fauna may have thrived in well-oxygenated, shallow, shelf-like waters close to the edge of the escarpment. Periodic severe storms may have generated turbulent currents which swept the animals and bottom sediment from the top of the cliff into deeper water through which they settled to the sea floor at the base of the cliff. There, the amount of dissolved oxygen in the water may have been very low, a so-called anoxic environment, wherein there would have been few if any bottom-scavenging creatures. As the animals settled to the bottom they quickly would have been buried by the fine muds raining down from atop the cliff. In this way the soft-bodied creatures could have been preserved, the carbon of their tissues being replaced, atom for atom, by compounds of calcium, silica and aluminum. The strata of the Burgess Shale show evidence that, if indeed they resulted from storm activity, many such storms occurred in the time interval during which the sediments and their enclosed fossils accumulated. For those interested in further pursuing this topic and enjoying the debate surrounding it, I have listed technical papers by Rolf Ludvigsen, Bill Fritz, and Jim Aitken and Ian McIlreath in Sources and Additional Reading.

The second issue to do with the Burgess Shale is the one addressed by Stephen Jay Gould in his book *Wonderful Life*. Drawing upon the exceptional work by Dr. Harry Whittington and his colleagues and students at Cambridge University, including Simon Conway Morris and Derek Briggs, Dr. Gould in essence attacks our conventional views of evolution, which describe a model of growing and expanding biological diversity throughout geological time that is capped by *Homo sapiens*. Rather than starting from simple beginnings during the Early Cambrian when animals began evolving into increasingly complex and diverse body designs, Dr. Gould describes a sudden and explosive Cambrian appearance of widely diverse and complex marine organisms, at least as diverse as the total of that in existence in our oceans today. Moreover, many of the groups present in the Burgess Shale did not survive beyond Cambrian time, having been "selected" out of existence by processes or means that were utterly capricious and unpredictable. Dr. Gould further asserts that the contingency, or unpredictability factor, acting upon an initially diverse fauna and its descendants throughout some 550 million years, was such as to so unpredictably modify the total mix of animal groups as to make the likelihood of the appearance of human beings most improbable. In other words, if one could replay the evolutionary tape many times, each replay of the tape would yield an entirely different set of animal survivors; the fact that primates appeared at all and that we evolved from one "pithicus" to another, until one of us ultimately became Sophia Loren, was pure fluke. I hope this "criminally" brief precis of current research into the fascinating world of the Burgess Shale and its fauna will entice you to read further.

Prior to and during the writing of this book scientists working in Newfoundland and Siberia recalibrated the geological clock and found that the Cambrian Period began 545 million years ago, rather than 570 million years ago as I have used elsewhere and shown on the geological column. The interval of time between about 600 million years and 545 million years has been called the Vendian Period of latest Precambrian time. It embraces the so-called "Ediacaran fauna," consisting of impressions of primitive jellyfish-like creatures and "**trace fossils**," a term for a range of horizontal tracks and burrows formed by worm-like animals. Thus, the length of the Cambrian period has been shortened considerably, from 65 million years to 40 million years (ending 505 million years ago). It was during the first twenty million years of this interval when virtually all of the major groups of animals, or **phyla**, first appeared, including representatives of the phylum Chordata, which contains humans. It has further been shown that this explosive rate of evolution was greatest between 530 million and 525 million years ago, just when the faunas of the Burgess Shale were thriving. Whereas most evolutionary concepts embody many tens of millions of years, the sudden appearence of multicellular life during the Early Cambrian is more akin to a biological "Big Bang." Moreover, discoveries of similar fossils in Greenland, China, Siberia and Namibia show that this Cambrian explosion of life occurred at virtually the same instant all around the world.

The cause of the biological explosion during the Early Cambrian is currently the focus of such frenzied research that by the time this book is published, ideas will have radically changed, more discoveries will have been made and dozens of scientific papers written on the subject. The current consensus among researchers is that the answer lies within the rocks deposited during the Vendian Period, the last twenty million years or so of the Precambrian Eon. Recent discoveries in Namibia have revealed the existence of a diverse community of strange fossil organisms that flourished in the oceans near the end of the Vendian Period. Considerable debate has ensued, however, on the relationship of these creatures to the modern animal kingdom and whether they, like the Ediacaran fauna, are evolutionary deadends or

From high on the slope of Mount Stephen the Kicking Horse River is seen to braid its course across a broad flood plain formed of river-transported sediment and landslide debris. (Photo by Lee McKenzie McAnally.)

our ancestors. Regardless of the outcome of this debate, what seems to have been needed to generate the biological Big Bang was the infusion of large amounts of oxygen into the oceans such that small, primitive creatures that had originated as far back as three-and-a-half billion years ago, could suddenly evolve into larger, more complex multicellular organisms during the Early Cambrian. Whereas throughout most of Precambrian time the rate of photosynthetic production of oxygen appears to have been balanced by oxygen-depleting processes such as organic decay, it is believed that about 600 million years ago, the rate of organic decay dramatically declined thus allowing significant quantities of oxygen to become available for rapid evolutionary processes to begin. Of course, another important ingredient needed was a sufficiently large tool kit of genes to work with. Some have suggested that about 550 million years ago, a worm-like creature was able to increase its critical number of genes from four to six or more, thus crossing some kind of genetic threshold and BOOM! – the Cambrian explosion.

The Kicking Horse River, from where it is joined by the Yoho River close to the base of the big hill to where it begins to carve its canyon down to the Columbia River, flows across a flat valley floor of composite origin. East of Field, the valley has been filled largely with

The Kicking Horse River rushes beneath a natural bridge eroded and dissolved out of carbonate strata of the Chancellor Formation.

extensive debris-flow deposits induced by jokulhlaups from Cathedral Glacier atop Cathedral Mountain which forms the southeast side of the valley (*see photo page 44*). From 1925, when historic debris-flows were first recorded, to 1984, some seven major events, with estimated slide volumes ranging between 5,000 and 136,000 cubic metres, have occurred. It is probable that such debris-flows have substantially filled the upper reaches of the Kicking Horse River valley, thus reducing the original gradient of the river and resulting in a braided stream pattern. However, to the south of the junction of the Kicking Horse and Emerald rivers, the floor of the broad, braided Kicking Horse River valley is filled mainly with glacial outwash that originated from melting glaciers in the Emerald, Amiskwi and Ottertail river valleys during Pleistocene deglaciation.

EMERALD LAKE AND THE NATURAL BRIDGE

A short distance past the entrance to Field is the intersection with the road to Emerald Lake. On your way to the lake you can stop and look at a natural bridge over the raging waters of the Kicking Horse River. The rocks here are steeply-inclined shaly limestone strata of the middle part of the Cambrian Chancellor Formation. At one time the river

Mount Burgess, to the south of Wapta Mountain, consists of shaly carbonate of the Lower Cambrian Chancellor Formation. The change in facies from Eldon carbonate to Chancellor limy shales approximately coincides with Burgess Pass between Wapta Mountain and Mount Burgess.

flowed as a waterfall across these rocks, however, continued erosion ultimately resulted in a breach in the base of the rock wall and the temporary formation of the bridge. It will not be long before the carbonate bridge succumbs to erosion and dissolution, leading to the eventual creation of a narrow gorge through the wall.

The origin of Emerald Lake is due to a combination of two or more glacial processes when a large glacier complex had developed on the President Range to the northwest and on other surrounding peaks such as Emerald Peak, Mount Carnarvon and Wapta Mountain. It is likely that glaciers flowing into Emerald River valley from these sources carved the large amphitheatre that includes the tarn basin holding the lake. It is likely that a recessional moraine also contributes to its enclosure.

The mountains surrounding the lake are formed from several formations of Cambrian age which have been disrupted by several northwesterly-trending thrust and normal faults. On Emerald Peak to the northwest of the lake, the Eldon and Pika formations, composed of carbonate and forming the northern flank of the mountain, change in facies to the dark, cleaved shaly carbonates and limy shale of the lower part of the Chancellor Formation on the southwest side of the mountain. The locus, or "line," across which the facies change occurs passes beneath the docks and bridge at the south end of the lake. From this location you can more clearly appreciate the facies change by looking upwards at Wapta Mountain (*see photo on page 70*) to the northeast and Mount Burgess to the southeast. The former peak is sculpted from Eldon Formation carbonate whereas Mount Burgess consists of Chancellor Formation muddy limestone strata. These two formations are the same age and occur on opposite sides of the Kicking Horse Rim (*see page 18*). By letting your eyes wander into the high valley between these mountains you can see into the western portal of Burgess Pass lying beneath Fossil Ridge which extends southeasterly from Wapta Mountain. High on the ridge is the quarry where Charles Walcott discovered the fossils within the Burgess Shale in the Stephen Formation (*see page 18*).

LARGE FOLDS IN THE MAIN RANGES

Upon returning to the Trans-Canada Highway and continuing westward you will notice that the geology appears to become increasingly less ordered. The rocks remain dark in colour and the structural style increases in complexity. These limy shales and muddy limestones of the Chancellor Formation were less able to withstand mountain-building forces than were the thick, resistant carbonates of the eastern Main Ranges and Front Ranges; consequently they, together with the Ottertail Formation carbonates, are deformed into many synclines and anticlines which are associated with a much larger feature called the Porcupine Creek **Anticlinorium** (*see page 110*). Whereas an anticline consists of a single upfold, an anticlinorium is a much larger upward flexure upon which many smaller satellite synclines and anticlines are superimposed. The opposite of an anticlinorium is a **synclinorium**, an example of which, called the Split Creek Synclinorium, occurs northwest of the bridge over the Ottertail River.

The core of the Porcupine Creek Anticlinorium occurs a few kilometres south of the bridge across the Ottertail River. Roadside outcrops of the Chancellor Formation consist of finely fractured slate which has sustained multiple phases of deformation during creation of the anticlinorium. In the drawing on page 109 I attempt to show the several effects of deformation upon these kinds of rocks.

Farther west, on the west side of the first bridge spanning the upper canyon of the Kicking Horse River, a thrust fault on the west flank of the anticlinorium has been rotated such that it and Ordovician McKay Group limestone strata on either side of the fault are now upside down. It should be pointed out, however, that the relative motion of the rocks on either side of the steeply-inclined fault, as shown by the single-barbed arrows in the photograph, is suggestive of a normal fault. In fact, on many field trips to this location I have heard geologists vigorously argue over just what kind of a fault this really is. It should be said that whenever you have seven geologists looking at

On the west side of the first bridge across the Kicking Horse River canyon, west of Field, steeply inclined strata of the Ordovician McKay Group are dislocated by a thrust fault which has been rotated into an overturned position on the western flank of the Porcupine Creek Anticlinorium (*see illustration on page 110*). The two, single barbed arrows on either side of the fault indicate the relative motion of one side with respect to the other. (Photo by W.R. Price.)

rocks you hear at least eleven opinions. The accompanying schematic cross-section by Hugh Balkwill (*see page 110*) illustrates the geometry of the anticlinorium and shows the locations of the photo and drawing, respectively on this page and page 109.

In its effort to reach the level of the Columbia River the Kicking Horse River has cut its canyon through the Beaverfoot Range, consisting of Ordovician clastic and lesser carbonate strata of the McKay Group as well as the Glenogle, Mount Wilson and Beaverfoot formations. This part of the river's course is in marked contrast with its upper reaches, where it braids its way across gravel valley-fill provided by numerous debris flows from the steep adjacent slopes as well as by glacial outwash. Whereas the upper valley was undoubtedly glaciated and widened, it is obvious that the lower canyon of the

Close to the centre, or core, of the Porcupine Creek Anticlinorium, my son Mark (about 1.2 m high) stands beside Cambrian strata of the Chancellor Formation which exhibit the effects of several stages of deformation that resulted in different kinds of structures including: the prominent, gently inclined cleavage, vertical fracturing, injection of an igneous dyke and its stretching into separate masses, or **boudins** (D). The stippled pattern represents the stratigraphic layering in the outcrop. The location of this drawing within the anticlinorium is shown in the illustration on page 110. Drawing by Eric Yorath based upon a photo by W.R. Price.

river was largely unaffected by glacial ice. The origin of the lower canyon is probably due to post-glacial headward erosion since the time when the valley was left hanging following retreat of the valley glacier that occupied the Columbia River valley in the Southern Rocky Mountain Trench.

ROCKY MOUNTAIN TRENCH

From the viewpoint above the town of Golden you get an excellent look at the Columbia River where it drains the Southern Rocky Mountain Trench. Across the valley is the Dogtooth Range of the Purcell Mountains. From Golden the Columbia River flows northward to Kinbasket Lake, where it turns southward to flow through Revelstoke and thence onward into the United States.

The Southern Rocky Mountain Trench, forming the western boundary of the Rockies, is part of a system of narrow, linear fault-controlled

Position of photo on page **108** —   — Position of drawing on page **109**

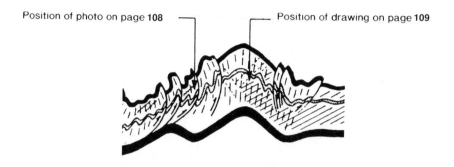

This diagramatic cross-section shows the internal structure of the Porcupine Creek Anticlinorium as interpreted along the Trans-Canada Highway. The stippled and two black layers represent the geometry of strata at different levels in the anticlinorium. The inclined dashed lines represent an early formed cleavage, whereas the vertical dashed lines show a later cleavage, each of which can be seen in the drawing on page 109. Based upon a drawing by H.R. Balkwill.

valleys extending the entire length of the Canadian Cordillera, a distance of about 3,500 kilometres. As such it is the longest valley system in the world and, according to American astronauts, is clearly visible from the moon. The northern two segments of the system, the Northern Rocky Mountain Trench in northeastern British Columbia and the Tintina Trench in the Yukon, are coincident with the surface traces of linear faults, called strike-slip faults, along which parts of the earth's crust have moved horizontally past one another. In these cases the block of crust on the west side of the fault has been displaced, perhaps as much as 750 kilometres northward relative to the block on the east side. Although more or less in line with the Northern Rocky Mountain Trench, the Southern Trench is not known to be coincident with a strike-slip fault but rather appears to have been partly formed along normal faults. In the vicinity of Golden, however, the trench appears to be due to erosion of soft shales in the core of a synclinorium underlying the floor of the valley.

The Icefields Parkway from Lake Louise to the Columbia Icefields

Unlike the route followed by the Trans-Canada Highway which takes you across the several ranges of the Rockies, the Icefields Parkway (Hwy 93) follows a course entirely within and parallel to the eastern Main Ranges. Consequently the geology remains more or less unchanged along the route, although several of the lower Paleozoic formations increase substantially in thickness. For the most part the route follows the surface trace of the Simpson Pass Thrust, which extends for some 320 kilometres from Mount Assiniboine in the south to near Jasper in the north.

The rocks along the Icefields Parkway, close to its junction with the Trans-Canada, consist of steeply-inclined cleaved slates, siltstones and sandstones of the Precambrian Corral Creek and Hector formations of the Miette Group. These strata form parts of nearly upright folds within the Simpson Pass thrust sheet; the surface trace of the Simpson Pass Thrust lies about one kilometre to the northeast of the highway.

HECTOR LAKE VIEWPOINT

From the viewpoint above Hector Lake (*see photo on page 69*), as at many other vantage points along this highway, you get an excellent view of the uniform layering, or stratification, in the eastern Main Ranges, as well as a glimpse of part of the Waputik Icefield which spawns several glaciers along the Continental Divide. Visible in the imposing cliffs beyond the lake are the Gog, Mount Whyte, Cathedral, Stephen and Eldon formations, all of Cambrian age. Carved from these same formations is Mount Hector to the east of the highway. This mountain, the lake and a nearby creek are named for Dr. James Hector, the medical doctor and naturalist attached to the Palliser expedition in the mid-1800s (*see page 6*).

Hector Lake recently has become the focus of research into past climates by Mel Reasoner, who conducted similar research on sediments recovered from Lake O'Hara as well as Bow and Crowfoot lakes (*see pages 98 and 115*). Together with a colleague, Eric Leonard, Mel recovered a core of sediment, 5.15 metres in length, from beneath the bottom of Hector Lake. The interesting thing about this core is that it is finely banded in alternating light and dark coloured layers averaging about one millimetre in thickness. These layers are called **varves** and result from the annual accumulation of silt due to inflow of sediment-laden, glacial meltwater during the summer (light coloured layer) and the slow settling of suspended clay and organic material during the winter when the lake is covered with ice and when inflowing streams are frozen (dark coloured layer). This kind of sedimentation is common in glacial lake settings. By counting the number of dark and light couplets in cores or exposed bluffs of ancient glacial strata, geologists can get extremely accurate measures of time between given events. The core from Hector Lake contains a record of 4,500 years of continuous varve sedimentation which continues today. Near the base of the core, the oldest couplet accumulated about 2,500 BC, just about the time when civilization was becoming established in the Indus Valley of Asia and at about the height of development of the "Old Kingdom" of Egypt. Since then the Hector Lake clock has kept ticking away, annually accumulating its pairs of varves: when the Greeks held the pass at Thermopylae in 480 BC (= 2477 couplets down from the top of the core), when Pontius Pilate washed his hands in Jerusalem in 36 AD (= 1961 couplets down), and when Albert Einstein published his Theory of Relativity in 1919 AD (= 78 couplets down). With luck, a couplet will, in future, record a year of peace throughout the world (???? AD).

This blocky outcrop along the highway opposite the south end of Bow Lake is one of the few known occurrences of igneous rocks in Banff National Park. The rocks, called diabase, form a dyke which has been injected into Precambrian Miette Group strata.

BOW LAKE AND CROWFOOT GLACIER

The area of the eastern Main Ranges embracing Bow Lake and Bow Summit as well as the line of impressive peaks between there and Saskatchewan Crossing displays glaciers, icefields and blue-green glacial lakes in settings unsurpassed for visual splendour along any other alpine highway in the world. As explained on page 56 the east-facing, steep slopes of the ranges often are mantled with glaciers and serrated by cirques due to the combined effects of air temperature, wind direction and physiography, whereas the west-facing slopes are relatively free of ice and snow. This phenomenon is particularly noticeable along this part of the Icefields Parkway where several tongues of ice, such as the Crowfoot, Bow and Peyto glaciers, drain the broad Wapta and Waputik icefields which straddle the Continental Divide in this area.

As you approach Crowfoot and Bow lakes from the south, keep an eye open for a dark rubbly outcrop on the east (right) side of the highway, immediately opposite the narrow connection between the two lakes. This is one of the very few known occurrences of igneous rocks in Banff and Yoho national parks. In this case the rock is a vertical sheet, or **dyke**, of diabase that has been injected, or intruded, into Precambrian strata of the Miette Group prior to the beginning of the Paleozoic Era; the dyke does not intrude the unconformably overlying strata of the Cambrian Gog Group. Diabase is a crystalline

The emerald-green waters of Bow Lake, seen from the Icefields Parkway, form the headwaters of the Bow River which drains southeastward beneath the Crowfoot Glacier on Crowfoot Mountain.

igneous rock composed of interlocking crystals of calcium-rich feldspar and **pyroxene**, minerals composed of various combinations of calcium, magnesium, aluminum, silica and oxygen. While in a fluid, mobile state, the molten magma was injected into the Miette strata, possibly during the time when North America (Laurentia) was splitting away from Australia and East Antarctica some 750 million years ago (*see page 15*). Another locality of igneous rocks is in southern Yoho National Park in the vicinity of the Ice River (*see geological map on page 58*). The Ice River Complex, as it is called, is a crystalline mass, or **pluton**, of Devonian age that is composed of a wide variety of minerals composed of alkalic elements (sodium, calcium, potassium) with gooey names like jacupirangite, ijolite and urtite.

Bow Lake is the source of the Bow River, or, more accurately, the river's headwaters are seen as Bow Glacier Falls which drain into the lake from Bow Glacier atop the prominent cliff in the background. The lake is enclosed within a glacial outwash plain which developed during the recessional stages of the last ice advance in this region. The rocks forming the many prominent cliffs are carbonates of Cambrian age which have been dislocated by the Cataract Brook Fault, a normal fault extending along the Continental Divide from the Lake O'Hara valley to the south (*see page 96*). To the south of the lake, Crowfoot Glacier mantles Cathedral Formation carbonates, the form of the ice mass resembling a crow's foot with one toe missing. A fresh moraine system occurs beneath the glacier close to Crowfoot Lake. This moraine complex as well as the lake sediments of Bow and Crowfoot lakes has also been the focus of Mel Reasoner's attention, where he has recognized the effects of a period of post-Wisconsinan global cooling followed by a dramatic period of warming. Based upon radiocarbon studies of sediment cores collected from Crowfoot and Bow lakes, as well as Lake O'Hara, Mel identified the moraines of the so-called "Crowfoot Advance" as having developed during an episode of cooling called the "Younger Dryas." It has long been known that the Younger Dryas cooling event occurred widely throughout Europe and the North Atlantic region over 10,000 years ago, during which time alpine glaciers advanced down their valleys. The recognition of this same event in western Canada indicates that the Younger Dryas cooling was a global phenomenon. The same cores also suggest that the Crowfoot Advance was abruptly terminated by a dramatic episode of warming. On the basis of measurements of the temperature-dependent ratio between the two isotopes of oxygen, oxygen 16 and oxygen 18 in Greenland ice cores and oceanic sediment cores, it is possible to calculate changes in temperature through time. Based upon these measurements and other evidence, it is thought that the warming event that terminated the Younger Dryas, about 10,100 years ago, resulted in an increase in global atmospheric temperatures of as much as seven degrees Celcius which may have occurred over a period of between only three and twenty years.

What could cause such dramatic warming? Oceanographers believe that such an event might be due to a sudden release of an enormous amount of heat stored in the oceans, possibly resulting from rapid changes in oceanic circulation. Another possible cause may have been a release of huge volumes of carbon monoxide into the atmosphere. This gas, a "greenhouse gas," is vastly more effective in enclosing radiated heat than is carbon dioxide, to which it quickly oxidizes. Large volumes of biogenic carbon monoxide, resulting from the decay of plants and animals, are trapped in arctic permafrost regions and within oceanic sediments at subduction zones where the earth's tectonic plates are converging upon one another and where the gas is in the form of frozen gas-hydrate, known as **clathrate**. Any persistent change in the temperature/pressure environment at which the clathrate is stable could cause substantial melting and release of the gas into the atmosphere. From several studies it is known that during the past 15,000 years the concentration of methane in the atmosphere has doubled and approaches the concentration level of

120,000 years ago, some 40,000 years prior to the onset of Wisconsinan glaciation. If such gas releases, together with escape of oceanic heat as a consequence of a sudden change in circulation patterns, were the causes of the dramatic increase in atmospheric temperature about 10,000 years ago, then we might take little comfort in the knowledge that these are natural processes about which we can do nothing if they should become active again. Together with the effects of Milankovitch cycles, which I discussed on page 40 and in the Glossary, it seems that dramatic changes in climate are the norm, rather than the exception.

PEYTO LAKE VIEWPOINT

At Bow Summit (elevation 2,070 metres) turn west along the short road leading to the parking lot from which a trail leads to the viewpoint above the southeast end of Peyto Lake. From here you get a magnificent view of Peyto Lake and the Mistaya River valley, which is developed along the surface trace of the Simpson Pass Thrust (*see photo page 72*).

The prominent mountain on the west side of Peyto Lake is Mount Mistaya. Here, Cambrian rocks form a westerly-inclined succession of strata in the hanging wall of the Simpson Pass thrust sheet. The lower tree-covered slopes are sandstones of the Gog Group, above which dominantly carbonate and lesser clastic strata of the Mount Whyte, Cathedral, Stephen, Eldon and Pika formations form a series of steep slopes. In the western distance, in front of Peyto Glacier, the Arctomys, Waterfowl, Sullivan and Lyell formations complete the thick Cambrian succession in this region.

The view to the northwest along the Mistaya River gives the impression that the valley has been carved into the core of an anticline with gently easterly-inclined strata on the east side of the valley and westerly-dipping strata to the west. Rather than forming an anticline, however, the strata on opposite sides of the valley belong to two, oppositely inclined and different thrust sheets. To the west the strata of Mount Mistaya and neighbouring peaks form part of the Simpson Pass thrust sheet; the strata to the east are part of the Pipestone Pass thrust sheet, which lies beneath the former and which exposes increasingly younger strata towards the north. In the far distance is the massive edifice of Mount Murchison, which occurs within the Pipestone Pass thrust sheet and exposes Upper Cambrian and Ordovician strata at the same level as Middle Cambrian strata on Mount Sarbach in the Simpson Pass thrust sheet on the opposite side of Mistaya Valley. Mount Murchison was named by James Hector for Sir Roderick Impy

This narrow canyon on the Mistaya River, above where it joins the North Saskatchewan, is carved into Cambrian carbonate strata of the Eldon Formation. Potholes, formed at earlier stages of canyon development, occur at different levels along the walls.

Murchison, director of the Geological Survey of Great Britain during the mid-nineteenth century and founding president of the prestigious Royal Society (London).

BOW SUMMIT TO SASKATCHEWAN CROSSING

Northward toward Saskatchewan Crossing the rugged beauty of the scenery continues. You pass by Mistaya Lake and the Waterfowl Lakes, all constrained by glacial moraine ridges, and above which stands the impressive horn peak of Mount Chephren, carved from quartzite strata of the Cambrian Gog Group (*see photo page 67*). At the crest of the hill above the Saskatchewan valley a trail leads down to Mistaya Canyon, a stunning potholed gorge cut through Cambrian carbonates by the raging Mistaya River as it descends to join the Saskatchewan River at Saskatchewan Crossing.

At Saskatchewan Crossing the Icefields Parkway crosses the North Saskatchewan River close to where it is joined by Howse River, Glacier River, and Arctomys Creek, three of several streams draining from the Lyell, Mons and Freshfield icefields adjacent to the Continental Divide. From a viewpoint just north of the bridge you can look out upon a family of peaks extending from the Mons Icefield in the west through mounts Forbes, Outram, Sarbach and Murchison to the east. Close by, the Icefields Parkway is joined by the David Thompson Highway, which leads eastward out of Banff National Park through spectacular mountain geology to beyond the mountain front.

MOUNT WILSON

Long before reaching Saskatchewan Crossing you get splendid views of the massive south face of Mount Wilson, named after Tom Wilson, a well-known guide and the first European to see Lake Louise (*see photo page 66*). The mountain is composed of several carbonate and shale formations of Late Cambrian and Early Ordovician age. The Upper Cambrian Waterfowl, Sullivan, Lyell and Bison Creek formations are largely tree-covered beneath the low but prominent cliff of Mistaya Formation carbonate, located about a third of the way up the mountain. Above the Mistaya are the Survey Peak, Outram, Skoki, Owen Creek, Mount Wilson and Beaverfoot formations, dominantly of Ordovician age; the last three of these are more fully developed on the adjacent peaks to the east where the prominent upper cliffs are formed of Mount Wilson quartzite. The boundary, or contact, between the Mount Wilson and the Beaverfoot Formation above it is unconformable, that is to say, it represents a gap in the geological record of about eight million years at this locality (*see page 22*).

Mount Wilson is an excellent mountain to illustrate what geologists call "**type sections**." When conducting field work, one of the many things we do is make detailed studies of the composition, internal structure and layering of rocks, a branch of geology called "stratigraphy." Part of the process is the recognition of mappable assemblages of strata which we call "formations." A formation is a body of strata of distinctive compositions (lithologies) which is recognizable throughout a broad area and which is readily distinguishable from other formations above and below. In other words it is a body of rock which geologists can recognize, wherever it occurs, and thus map its distribution. When defining a formation the geologist studies its composition, measures its thickness, collects its fossils, if any, and does similar studies on other exposures of the same strata at other places. Once having gathered sufficient data on the nature of the formation, he or she then chooses a name for the formation, ideally the name of a geographical feature, such as a river or mountain, that is close to where the formation is best and most completely exposed. This locality is designated as the formation's type section, with which

Shortly after descending from its source on the Saskatchewan Glacier, the North Saskatchewan River dumps its load of sediment to form broad gravel flats across which it braids its course alongside the Icefields Parkway.

all other exposures of the formation can be compared.

Mount Wilson exposes four type sections: the Survey Peak, Outram, Owen Creek and Mount Wilson formations. The names for the first three were obtained from nearby geographical features, because the name "Mount Wilson" had already been used by Charles Walcott in 1923, the same man who discovered the Burgess Shale (*see page 18*).

SASKATCHEWAN CROSSING TO SUNWAPTA PASS

Northward from Saskatchewan Crossing the valley of the North Saskatchewan River is broad, with extensive flats across which the river braids its course as it dumps its load of sand and gravel carried from its source at the snout of Saskatchewan Glacier. Close to the base of the Big Hill which takes you up to Sunwapta Pass and the Columbia Icefields, you pass by an imposing cliff of Devonian Palliser Formation carbonate known as the Weeping Wall, its name given for the several small springs cascading down its surface (*see photo page 26*). The Weeping Wall forms the lower cliff of Cirrus Mountain, the upper part of which exposes the other two formations of the well-known Front Ranges triplet, the Carboniferous Banff and Livingstone formations (*see photo page 72*). Ahead you can see a text-book example of a syncline formed by strata in the mountains closest to and farthest from you. On the left, the west limb of the syncline is expressed by strata of the Palliser Formation; to the right, in the distance, the remainder of the structure can be seen on Nigel Peak, the upper part of which is carved from the Banff and Livingstone formations (*see photo page 69*).

Cresting the Big Hill you pass by Parker Ridge on your left, a popular alpine hiking and cross-country ski area formed from the Devonian Southesk Formation that is part of the Fairholme Group. In front of you is the south face of Mount Athabasca, traversed by the Simpson Pass Thrust and situated adjacent to Sunwapta Pass, the boundary between Banff and Jasper national parks.

It is in this region where Canada's longest, and one of its deepest caves occurs, in carbonates of the Cambrian Cathedral Formation. Castleguard Cave consists of over thirteen km of passageways, ten of which comprise the main passage that extends beneath part of the Columbia Icefield on Castleguard Mountain. An outstanding and beautifully illustrated book on Castleguard Cave has been written by Derek Ford and D. Muir (*see Sources and Additional Reading*).

THE COLUMBIA ICEFIELD

The Columbia Icefield and neighbouring Chaba and Clemenceau icefields form a more or less continuous mass of ice along the Continental Divide in the vicinity of the boundary between Banff and Jasper national parks. This composite icefield is drained by several glaciers, meltwaters from which supply the Pacific, Arctic and Atlantic drainage systems, all of which diverge from the Snow Dome in the Columbia Icefield, the hydrographic apex of North America. For a full discussion of the Athabasca Glacier in Jasper National Park, the reader is referred to the book *Of Rocks, Mountains and Jasper* by myself and Ben Gadd (*see Sources and Additional Reading*).

Within the Columbia Icefield the boundary between Banff and Jasper national parks crosses the Snow Dome and Mount Athabasca. Between Mount Athabasca and Mount Saskatchewan, the Saskatchewan Glacier flows easterly from off the north slope of Castleguard Mountain and supplies the headwaters of the North Saskatchewan River.

Like many other glaciers in the Rockies, those draining the Columbia and neighbouring icefields sustained a period of substantial advance during the so-called "Little Ice Age." During an interval of some 900 years, beginning around 950 AD, global temperatures began to cool such that, by about 1500 winter snow accumulation exceeded summer melting, thus favouring the growth and advance of icefields and valley glaciers throughout Europe, Asia and North America. By around 1840 the Saskatchewan and Athabasca glaciers had reached their maximum down-valley extent and since then have been retreating; the current rate of meltback is between fifteen and twenty metres per year.

Will our glaciers continue to retreat indefinitely? The answer to this question is complex and uncertain. Most geologists would agree that, under normal, natural circumstances, we are still in the same glacial cycle that began over two million years ago and that we again could experience yet another hemispheric glaciation. On the other hand, many scientists predict that the effects of atmospheric pollution ultimately will result in global warming and the melting of our glaciers and ice caps. Currently there is much debate in the scientific community on the long-term effects of such pollution. Many believe that industrial and agricultural production of carbon dioxide, carbon monoxide and other gasses will result in a global increase in temperature of between two and five degrees Celsius within the next seventy-five years (the greenhouse effect), the effects of which would be substantial indeed. Glaciers would melt, resulting in a rise in sea level leading to the flooding of many coastal communities where most of the world's populations are located. Contrary to this idea, many scientists argue that the currently observed trend in increased average temperature may be a normal climatic variation following the Little Ice Age (*see page 89*) and, moreover, that we have insufficient data to predict the precise consequences of increased concentrations of greenhouse gasses. Whatever the ultimate truth may be, it is, I believe, immoral to beg the question. Even if we don't know the answer, as a self-proclaimed intelligent form of life, we should stop using the atmosphere as a receptacle for our gaseous garbage. Then, who knows, maybe some day our glaciers will grow again. On that note I bid you good day.

Sources and Additional Reading

Several of the references listed below are out of print. These, as well as others listed, can be examined in the libraries of the Geological Survey of Canada offices in Calgary, Vancouver and Ottawa and at the libraries of Canadian universities with geoscience departments.

Aitken, J.D. and McIlreath, I.A., 1990. "Comments and Reply on 'The Burgess Shale: Not in the Shadow of the Cathedral Escarpment'" by Rolf Ludvigsen; *Geoscience Canada*, 17, pp. 111-115.

Aitken, J.D., Fritz, W.H. and Norford, B.S., 1972. "Cambrian and Ordovician biostratigraphy of the southern Canadian Rocky Mountains"; *XXIV International Geological Congress*, Excursion A-19.

Akrigg, G.P.V and Akrigg, H.B., 1988. *British Columbia Place Names*; Sono Nis Press, Victoria; 346 pages.

Bally, A.W. (ed), 1989. *The Geology of North America: An Overview*; Geological Society of America, Boulder; 619 pages.

Belyea, H.R. and Labrecque, J.E., 1972. "Devonian stratigraphy and facies of the southern Rocky Mountains of Canada, and the adjacent plains"; *XXIV International Geological Congress*, Excursion C-18. Out of print.

Bobrowsky P. and Rutter, N.W., 1992. "The Quaternary geologic history of the Canadian Rocky Mountains"; *Geographie physique et Quaternaire*, 46, pp. 5-50.

Briggs, D.E.G., Erwin, D.H. and Collier, F.J., 1994. *The Fossils of the Burgess Shale*; Smithsonian Institution Press, Washington and London; 238 pages.

Conway Morria, S. and Whittington, H.B., 1985. *Fossils of the Burgess Shale*; Geological Survey of Canada, Miscellaneous Report 43, 31 pages.

Dodd, John and Helgason, Gail, 1991. *The Canadian Rockies Access Guide*; Lone Pine Publishing, Edmonton; 360 pages.

Douglas, R.J.W. (ed), 1970. *Geology and economic minerals of Canada*; Geological Survey of Canada, Economic Geology Report No. 1, 5th edition; 838 pages, maps, cross-sections, correlation charts.

Fritz, W.H., 1990. "In defence of the escarpment near the Burgess Shale fossil locality"; Comments and Reply on 'The Burgess Shale: Not in the Shadow of the Cathedral Escarpment' by Rolf Ludvigsen; *Geoscience Canada*, 17, pp. 106-110.

Gabrielse, H. and Yorath, C.J. (eds), 1991. "Geology of the Cordilleran Orogen in Canada"; Geological Survey of Canada, *Geology of Canada*, No. 4; 844 pages, maps, cross-sections, correlation charts.

Gadd, Ben, 1995. *Handbook of the Canadian Rockies*; Corax Publishing, Jasper; 831 pages.

Gould, S. J., 1989. *Wonderful Life: The Burgess Shale and the Nature of History*; W.W. Norton, New York; 347 pages.

Gretener, P.E., 1967. "Significance of the rare event in geology"; *The American Association of Petroleum Geologists Bulletin*, 51, pp. 2197-2206.

Halliday, I.A.R. and Mathewson, D.H. (eds), 1971. *A geological guide of the eastern Cordillera along the Trans-Canada Highway between Calgary and Revelstoke*; Canadian Society of Petroleum Geologists; 94 pages, maps, correlation charts.

Holmgren, E.J. and Holmgren, P., 1973. *Over 2,000 Place-Names of Alberta*; Western Producer Book Service, Saskatoon; 240 pages.

Jackson, L.E., Hunger, O., Gardner, J.S. and Mackay, C., 1989. "Cathedral mountain debris flows, Canada"; *Bulletin of the International Association of Engineering Geology*, No. 40, pp. 35-54.

Langshaw, Rick, 1989. *Geology of the Canadian Rockies*; Summerthought Ltd., Banff; 64 pages.

Ludvigsen, Rolf, 1989. "The Burgess Shale: Not in the Shadow of the Cathedral Escarpment"; *Geoscience Canada*, 16, pp. 51-59.

McGugan, A. and Rapson-McGugan, J.E., 1972. "The Permian of the southeastern Cordillera"; *XXIV International Geological Congress*, Excursion A-16. Out of print.

Muir, D. and Ford, D., 1985. *Castleguard*; Parks Canada, 244 pages.

Monger, J.W.H., Clowes, R.M., Cowan, D.S., Potter, C.J., Price, R.A. and Yorath, C.J., 1994. "Continent-ocean transitions in western North America between latitudes 46 and 56 degrees: Transects B1, B2, B3"; in *Phanerozoic evolution of North American continent-ocean transitions*, R.C. Speed (ed); Decade of North American Geology, Geological Society of America, pp. 357–397.

Mossop, G. and Shetsen, I. (eds), 1994. *Geological Atlas of the Western Canada Sedimentary Basin*; Canadian Society of Petroleum Geologists, Calgary, 510 pages.

Mountjoy, E.W. and Geldsetzer, H.H.J, 1981. "Devonian stratigraphy and sedimentation, southern Rocky Mountains"; in *Field Guides to Geology and Mineral Deposits, Calgary '81*, R.I. Thompson and D.G. Cook (eds); Geological Association of Canada/Mineralogical Association of Canada/Canadian Geophysical Union, Joint Annual Meeting, Calgary, pp. 195 – 224.

Patton, Brian and Robinson, Bart, 1990. *The Canadian Rockies Trail Guide*; Summerthought Ltd., Banff; 363 pages.

Price, R.A., 1981. "The Cordilleran foreland thrust and fold belt in the southern Canadian Rocky Mountains"; in *Thrust and Nappe Tectonics*; Geological Society of London, pp. 427–448.

Price, R.A., Balkwill, H.R., Charlesworth, H.A.K., Cook, D.G. and Simony, P.S., 1972. "The Canadian Rockies and Tectonic Evolution of the Southeastern Canadian Cordillera"; *XXIV International Geological Congress*, Field Excursion AC-15. Out of print.

Price, R.A., Monger, J.W.H. and Roddick, J.A., 1985. "Cordilleran cross-section: Calgary to Vancouver"; in *Field Guides to Geology and Mineral Deposits in the Canadian Cordillera*, D.J. Templeman-Kluit (ed); Geological Society of America, Cordilleran Section Meeting, Vancouver, 1985. pp. 3-1 – 3-85.

Proudfoot, D.N., Moran, S.R. and Rutter, N.W., 1981. "The Quaternary strati- graphy and geomorphology of southwest/central Alberta"; in *Field Guides to Geology and Mineral Deposits, Calgary '81*, R.I. Thompson and D.G. Cook (eds); Geological Association of Canada/Mineralogical Association of Canada/Canadian Geophysical Union Joint Annual Meeting, Calgary, 1981. pp. 225–259.

Rutter, N.W. and Christiansen, E.A., 1972. "Quaternary Geology and Geomorphology Between Winnipeg and the Rocky Mountains"; *XXIV International Geological Congress*, Field Excursion C-22. Out of print.

Stelck, C.R., Wall, J.H., Williams, G.D. and Mellon, G.B., 1972. "The

Cretaceous and Jurassic of the Foothills of the Rocky Mountains of Alberta"; *XXIV International Geological Congress*, Excursion A-20. Out of print.

Stott, D.F. and Aitken, J.D. (eds), 1993. "Sedimentary Cover of the Craton in Canada"; Geological Survey of Canada, *Geology of Canada*, No. 5; 826 pages.

Yorath, C.J., 1990. *Where Terranes Collide*; Orca Book Publishers, Victoria; 231 pages.

Yorath, Chris and Gadd, Ben, 1995. *Of Rocks, Mountains and Jasper*; Dundurn Press, Toronto; 170 pages.

Young, Grant M., 1991. "The geological record of glaciation: Relevance to the climatic history of the earth"; *Geoscience Canada*, 18, pp. 100–108.

Zaslow, Morris, 1975. *Reading the Rocks: The Story of the Geological Survey of Canada, 1842 - 1972*; Macmillan Company of Canada in association with the Geological Survey of Canada. Ottawa; 599 pages.

MAPS

Geological maps and other publications of the Geological Survey of Canada can be purchased from the following offices of the Geological Survey of Canada:

Geological Survey of Canada – Calgary
3303 - 33 St. N.W.
Calgary, AB. T2L 2A7
Publication Sales: 403-292-7030

Geological Survey of Canada – Vancouver
100 West Pender St.
Vancouver, BC. V6B 1R8
Publication Sales: 604-666-6840

Geological Survey of Canada – Ottawa
601 Booth St.
Ottawa, ON. K1A 0E8
Publication Sales: 613-952-8782

The following coloured geological maps illustrate much of the discussion in this guide; each is accompanied with cross-sections and a legend.

Balkwill, H.R., Price, R.A., Cook, D.G., and Mountjoy, E.W., 1980. Geology of Golden; Geological Survey of Canada, maps 1496A and 1497A; 1:50,000.

Geological Survey of Canada, 1978. Geology of Calgary; Geological Survey of Canada, map 1457A; 1:250,000.

Ollerenshaw, N.C., 1972. Geology of Lake Minnewanka; Geological Survey of Canada, maps 1347A (1271A) and 1272A; 1:50,000.

Price, R.A. and Mountjoy, E.W., 1970. Geology of Canmore; Geological Survey of Canada, maps 1265A and 1266A; 1:50,000.

Price R.A. and Mountjoy, E.W., 1972. Geology of Banff; Geological Survey of Canada, maps 1294A and 1295A; 1:50,000.

Price, R.A. and Mountjoy, E.W., 1972. Geology of Mount Eisenhower; Geological Survey of Canada, maps 1296A and 1297A; 1:50,000.

Price, R.A. and Mountjoy, E.W., 1978. Geology of Hector Lake; Geological Survey of Canada, maps 1463A and 1464A; 1:50,000.

Price, R.A., Cook, D.G., Aitken, J.D. and Mountjoy E.W., 1980. Geology of Lake Louise; Geological Survey of Canada, maps 1482A and 1483A; 1:50,000.

GLOSSARY

Alluvial fan. An outward spreading, fan-shaped mass of loose rock material deposited by: (1) a stream where it leaves a narrow mountain valley and flows across an adjacent plain, or (2) where a tributary stream of steep gradient joins a main stream of lower gradient.

Antecedent stream. A stream established prior to uplift of the mountain ranges through which it has cut its course.

Anticline. An up-fold in which the strata on either side of the crest have been bent downward. The oldest strata are found in the centre of the fold.

Anticlinorium. A large upward flexure with many superimposed satellite smaller anticlines and synclines.

Apatite. A group of minerals composed of calcium phosphate with varying amounts of fluorine and chlorine, formed in a variety of ways including precipitation from sea water.

Archaeocyathids. Cone-, goblet- or vase-shaped organisms related to sponges with a skeleton of calcium carbonate. Archaeocyathids lived in reef colonies throughout the world's oceans during the early part of the Cambrian Period.

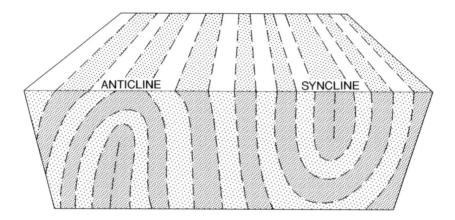

Arête. A narrow, serrated mountain ridge developed by the glacial sculpting of the headwalls of adjacent cirques.

Atoll. A roughly circular or horseshoe-shaped coral reef, enclosing a shallow lagoon and commonly constructed upon the tops of individual volcanoes of volcanic island arcs.

Base level. That level below which a stream cannot erode, usually sea level or the level of a large lake.

Basement. As used in this book a term referring to the Precambrian crystalline igneous and metamorphic rocks which extend from the Canadian Shield westward beneath the plains and Rocky Mountains.

Bed. See **stratum**.

Braided stream. One flowing in several dividing and reuniting channels resembling the strands of a braid. They occur where more sediment is being brought to any part of a stream than it can remove. The building of bars becomes excessive, forcing the stream to develop an intricate network of interlacing channels.

Calcite. Calcium carbonate ($CaCO_3$), the mineral that forms limestone.

Canadian Shield. The vast, shield-shaped region of exposed Precambrian crystalline rocks centred upon Hudson Bay. The region underlies more than five million square kilometres and is composed of welded fragments of crust formed between 4.0 and 2.5 billion years ago. Most of the world's continents consist of one or more Precambrian shields around and on top of which younger rocks have accumulated.

Carbonates. As used herein, carbonates are sedimentary rocks formed by the organic precipitation from marine waters of mineral carbonates of calcium and magnesium. Examples are limestone (calcium carbonate, the mineral calcite) and dolomite (calcium and magnesium carbonate).

Chert. A dense sedimentary rock consisting of extremely fine, interlocking crystals of quartz, occurring as nodules or beds within carbonates or as extensive layers resulting from the accumulation of radiolaria shells.

Chlorite. A hydrous greenish mineral, resembling mica, consisting of various combinations of magnesium, iron, aluminum and silica.

Cirque. A steep-walled, amphitheatre-like recess occurring at high elevations on the side of a mountain, commonly at the head of a glacial valley and formed from the erosive carving of a mountain glacier. Small lakes, called tarns, commonly occur at the base of cirque headwalls.

Clastics. Sedimentary rocks composed of broken fragments (clasts) derived through erosion of pre-existing rocks or minerals and transported by streams, wind, etc. to their place of deposition. The most common clastic rocks are quartz-sandstone and shale.

Clathrate. A naturally occurring solid comprised of water molecules forming a rigid lattice of atom-bounded cages, each containing a molecule of natural gas, usually methane. The water crystallizes in the isometric (cubic) crystallographic system, rather than the hexagonal system of normal ice.

Clay. Rocks or mineral fragments of any composition having diameters of less than 1/256 mm. The mineral fragments commonly belong to a complex and loosely defined group of hydrous silicates of aluminum, iron and magnesium.

Cleavage. The tendency of some rocks to cleave or split along parallel, closely spaced fracture planes which may be parallel to, or inclined to, the stratification. Cleavage is developed under conditions of increased pressure, commonly during episodes of mountain building.

Columbia Mountains. Occurring in southeastern British Columbia, the Columbia Mountains include the Cariboo, Selkirk and Purcell mountains.

Competence. (1) The ability of a stream to transport a given sediment load. (2) The ability of rocks, depending upon their strength and rigidity, to withstand deforming forces.

Conglomerate. A coarse-grained clastic sedimentary rock, composed of rounded to sub-angular fragments larger than 2 mm in diameter within a matrix of sand or silt (consolidated gravel).

Consequent stream. As used in this book, a stream flowing down the dip slope of a mountain range.

Continental Divide. A drainage divide separating streams flowing towards different oceans surrounding a continent. In the Banff and Yoho Rockies the Continental Divide separates streams flowing into the Pacific Ocean from those draining into the Atlantic Ocean (Hudson Bay).

Continental Drift. The theory that the continents were once joined into a single supercontinent (Pangea) which broke apart with the various fragments (continents) moving away from one another. As used today the term embraces the hypotheses of seafloor spreading and plate tectonics.

Cordillera. The system of mountain ranges, valleys and plateaus extending from the southern tip of South America to the Arctic Ocean. It forms the backbone or continental spine of western North America.

Cordilleran Ice Sheet. See **Wisconsinan glacial stage**.

Crust. The crust forms the outermost layer of the earth and is of two types. That which forms the continents is between 32 and about 55 kilometres thick and composed of minerals of mainly sodium, potassium, aluminum, silicon and oxygen. Oceanic crust, excluding its thin cover of sediments, is about 5 to 7 kilometres thick and made of compounds of dominantly iron, magnesium, silicon and oxygen.

Cyanobacteria. Blue-green algae.

Décollement. As used in this guide, the principal surface along which the sedimentary rocks forming the Rocky Mountains were detached from and moved eastwards over the Precambrian crystalline granitic rocks of the buried western flank of the Canadian Shield. The décollement forms a common zone wherein the several main thrust faults merge into a common surface of detachment.

Diamictites. As used in this guide, a rock containing a large variety of different types and sizes of clasts in a matrix of clay or mudstone and deposited upon the sea floor through the melting of glaciers or ice shelves.

Differential erosion. Erosion which occurs at varying rates due to differences in the composition of adjacent rock types. Given the climate in which they occur, resistant rocks form cliffs while more easily eroded recessive rocks form less steeply inclined slopes.

Dip. The angle of downward inclination from the horizontal of a planar surface such as a bedding surface, fault, etc. Dip slope is a surface slope parallel to and formed by the dip of strata.

Dolomite. See **carbonates**.

Dolomitization. The conversion of limestone to dolomite through the addition of magnesium. Dolomitization occurs shortly after deposition of limestone through the action of magnesium-bearing marine water.

Dyke. A tabular body of igneous rock which has been intruded into and at an angle to the stratigraphic layering of its host rocks.

Erratic. As used in this guide, a large boulder that has been carried to its resting place by glacial ice. Far-travelled erratics are usually recognized by their difference in composition from the bedrock upon which they rest.

Evaporites. Sedimentary rocks composed of minerals which precipitate from evaporating sea water.

Facies. As used in this guide, a general term embracing the composition and depositional environments of rocks. A facies change refers to a change in rock composition and inferred environment of deposition from one region to another during a given interval of time.

Feldspars. A group of closely related minerals composed of silica, oxygen, aluminum and one or more of potassium, sodium and calcium. The feldspars are the most abundant of all minerals, formed by crystallization from molten magmas. They are the most common constituents of igneous rocks and occur as mineral grains in sedimentary rocks.

Folds. See **anticline** and **syncline**.

Formation. Stratified rocks are given names according to many criteria, including their composition and location. The most fundamental type of subdivision is the formation which refers to a succession of strata with specific characteristics and which can be recognized as distinctly different from other formations throughout its region of occurrence. For example, the Corral Creek Formation is the name given to a succession of multi-coloured slates that occur widely throughout the Banff Rockies but which are best exposed along Corral Creek in the Slate Range near the Whitehorn ski development where they were first studied in detail. In some instances several formations sharing common origins or other characteristics can be referred collectively to a group; the Corral Creek Formation is one of several formations assigned to the Miette Group. There are supergroups too. The Miette Group is part of the Windermere Supergroup; however, I prefer to leave that one alone.

Foothills. The region of low, rounded hills fringing the easternmost range of the Rockies and bounded to the east by the flat to gently rolling plains.

Footwall. See **thrust fault**.

Front Ranges. Those northwesterly trending ranges occurring between the western limit of the foothills and the eastern edge of the main ranges. The Front Ranges consist mainly of Devonian and younger carbonate strata which occur in westerly inclined, thick thrust sheets, each range separated by linear valleys underlain by shales.

Garnets. A group of minerals consisting of various combinations of calcium, magnesium, iron, aluminum, manganese, vanadium, chromium, silica and oxygen. Although they occur as minor constituents of igneous rocks, they most commonly are found as distinctive red crystals in metamorphic rocks where they formed at high pressures and temperatures.

Geological Survey of Canada (GSC). Canada's largest scientific research organization and one of the oldest and most prestigious of its kind in the world. Since its establishment in 1842 the GSC has provided Canadians with an understanding of the geological architecture of the country, with a view to providing guidance in the responsible development and use of its mineral and energy resources as well as a knowledge of its natural hazards. With headquarters in Ottawa, it has regional offices in Dartmouth, Nova Scotia, Quebec City, Calgary, Alberta and in Vancouver and Victoria, British Columbia.

Geothermal gradient. The increase in temperature with depth. Near the earth's surface the rate of increase is about 25 degrees Celsius per kilometre but varies with the type of crust (continental or oceanic) and tectonic setting (mountains versus plains etc).

Glacier. A large mass of long-lasting ice formed, on land, by the compaction and recrystallization of snow, moving slowly by creep downslope, or outward in all directions due to the stress of its own weight.

Glacial rebound. Uplift of the crust that occurs after the retreat of a continental glacier, in response to earlier subsidence under the weight of the ice. See **isostasy**.

Glaciation. The formation, movement and recession of ice sheets. A collective term for the geological processes associated with glacial activity, including erosion and deposition, and the resulting effects of such action on the earth's surface.

Gneiss. See **metamorphic rocks**.

Gondwanaland. See **Pangea**.

Graptolites. Tiny colonial floating organisms which lived throughout the Paleozoic Era. The organisms consisted of a rigid "spine" along which were arranged individual living chambers occupied by a polyp-like organism. Graptolites are commonly found on the bedding surfaces of black shales where they resemble narrow leaves.

Group. See **Formation**

Hanging valley. A tributary glacial valley, the mouth of which is at a relatively high elevation on the steep side of a larger main glacial valley.

Hanging wall. See **thrust fault**.

Hoodoos. Pillars developed by rainfall runoff erosion of semi-consolidated horizontal sedimentary strata of varying resistance.

Horn peak. A high, rocky, sharp-pointed mountain peak with prominent faces and ridges, bounded by the intersecting walls of three or more cirques that have been cut by the headward erosion of mountain glaciers.

Icefield. An extensive mass of land ice covering a mountain region, consisting of many interconnected alpine glaciers and covering all but the highest peaks and ridges.

Ingenous rocks. Rocks which have formed by cooling from a molten state (ie., lava, granite, etc.)

Insular Superterrane. See **Superterranes**.

Intermontane Superterrane. See **Superterranes**.

Island arcs. Chains of volcanoes constructed above subducting oceanic crust. As oceanic crust is subducted beneath another plate it reaches depths where the temperature is sufficient to melt the subducting plate. From there molten material rises upward through the overriding plate to appear at the surface as a chain of volcanoes. In the oceans these chains are called volcanic island arcs and on land they are called volcanic arcs.

Isostasy. A condition analogous to floating whereby, when a load is placed upon the earth's crust, the crust sinks proportionally to the mass of the applied load.

Joints. Fractures, commonly occurring in two or more sets, on either side of which there has been no appreciable movement of the rock.

Jokulhlaup. An Icelandic word for the sudden release of glacier-entrained water, either a lake or water-filled ice cavern.

Karst. A type of topography that is formed by the dissolving action of water on carbonates and characterized by sinkholes, caves or underground drainage.

Kicking Horse Rim. A physiographic feature which existed during early Paleozoic time, near the edge of the continent and which separated shallow waters of the miogeocline from deeper waters covering the edge of the continent. Near Field, B.C. it appears to have been a steep, submarine cliff of carbonate belonging to the Cambrian Cathedral Formation and called the Cathedral Escarpment. The Kicking Horse Rim coincides with the facies change from eastern carbonates to western limy shales.

Lateral moraine. See **moraine**.

Laurentia. A large Precambrian land mass consisting of the Canadian Shield, the Greenland Shield, the Baltic Shield and the Precambrian shields of Antarctica, Australia and Siberia.

Laurentide Ice Sheet. See **Wisconsinan glacial stage**.

Laurasia. See **Pangea**.

Limb (of a fold). One side of an **anticline** or **syncline**.

Limestone. See **carbonates**.

Lithification. The processes by which sediments such as gravel, sand and mud are converted to rocks such as conglomerate, sandstone and shale.

Lithology. The composition of rocks.

Little Ice Age. A period within the past several centuries during which mountain glaciers attained their maximum extent. In the Rockies glaciers began their advance (Cavell Advance) probably prior to 600 years ago (1400 AD) and reached their maximum extents between 100 and 400 years ago.

Loess. A loose deposit of silt and clay deposited by the wind.

Magma. Molten rock from which igneous rocks and minerals form by cooling and solidification. Most magmas originate in the mantle from where they are injected into the overlying crust.

Main ranges. Those northwesterly trending ranges of the Rockies consisting mainly of Cambrian and Precambrian strata that are deformed into complex folds and broken by thrust and normal faults.

Mantle. The thick layer between the base of the earth's crust and the outer core, composed of compounds largely of iron, magnesium, silicon and oxygen.

Metamorphic rocks. Rocks which, since their initial formation, have sustained increases in temperature and pressure sufficient to change their original minerals and textures to new minerals and new textures. **Slate** is the metamorphic equivalent of shale and can be split into thin slabs. Schist is a foliated, crystalline rock that can be split into slabs due to the parallelism of the minerals present. Gneiss is a foliated rock which is commonly banded due to alternating layers of dark and light coloured minerals. It does not readily split into slabs.

Milankovitch cycles. In the early 1920s Milutin Milankovitch proposed that periodic irregularities in the earth's rotational and orbital characteristics are sufficient to alter the amount of solar radiation received by the earth at a given latitude, thus leading to changes in climate. Maximum changes in orbital eccentricity, the degree to which the earth's orbit departs from a perfect circle, occur about every 100,000 years when the earth is farthest from the sun. Also, the angle between the earth's rotational axis and the orbital plane, currently about 22.5 degrees, shifts by about 1.5 degrees every 41,000 years. A third cycle is the precession of the equinoxes, or the two orbital positions where the the sun crosses the equator (March 21 and September 21); these positions slowly shift around the orbit in a 23,000-year cycle. Continuous

Incipient normal fault

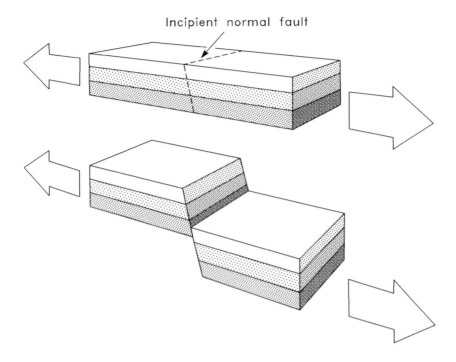

changes in these three characteristics cause the amount of solar heat received to vary over time and provide the triggering mechanism for glaciations.

Miogeocline. As used in this book, a term embracing both the westward-sloping shelf of the continent from late Precambrian to Middle Jurassic time and the sediments which accumulated upon it during that 600 million-year interval.

Moraine. A ridge of unsorted gravel, sand and silt deposited by an active glacier. There are several types including lateral moraines deposited along the sides of a glacier, and terminal and recessional moraines respectively deposited at the farthest point of advance of a glacier and in stages of still-stand during retreat of a glacier.

Normal fault. A steep fault, usually inclined at an angle greater than 45 degrees, along which the rocks above the fault have moved downward, relative to those beneath the fault. Whereas thrust faults result from compressive forces, normal faults occur when the rocks are subjected to tensile forces.

Omineca Belt. One of five subdivisions of the Canadian Cordillera, differentiated from one another by a combination of rock types, geological history and physiography. The easternmost of these, containing the Rockies, is the "Foreland Belt," consisting entirely of sedimentary rocks. To the west, the "Omineca Belt" which includes the Purcell, Selkirk, Monashee, Cariboo, Omineca, Cassiar and Selwyn mountains, consists of sedimentary, igneous and metamorphic rocks. The "Intermontane Belt" in the central part of the Cordillera is dominated by volcanic rocks and forms the broad interior plateaus of British Columbia and the Yukon, but also contains the rugged Skeena Mountains of north-central British Columbia. To the west the "Coast Belt" consists of rugged peaks of granitic igneous rocks. Farthest west is the "Insular Belt," formed of sedimentary and igneous rocks making up Vancouver Island, the Queen Charlotte Islands as well as the Saint Elias Mountains, the latter being the highest and most rugged mountains in North America.

Pangea (also **Pangaea**). The name given to a giant supercontinent, including all of the world's continents, which had assembled by about 300 million years ago and which began to break up about 200 million years ago. The initial break-up resulted in the separation of Laurasia including North America and Eurasia, from Gondwanaland, which contained South America, Africa, India, Australia and Antarctica.

Panthalasa. The globe-encircling ocean surrounding Pangea, or, a men's wear shop in Lhasa, the capital of Tibet.

Phosphorite. See **apatite**.

Phylum. A major grouping of organisms (below kingdom and sub-kingdom) sharing similar patterns of organization and a line of evolution from a presumed common ancestor.

Physiography. The description of the earth's surface features and landforms.

Plate tectonics. The corollary hypothesis to seafloor spreading which explains the motions of the world's crustal plates and the subduction of oceanic crust along the world's deep sea trenches.

Platform. (1) A broad bank of carbonate. (2) A general term for the surface of the interior of a continent; in this sense platformal rocks are those which accumulated east of the miogeocline.

Pleistocene Epoch. See **Quaternary Period**.

Pluton. A body of coarsely crystalline igneous rock originating from solidification of molten magma intruded into the crust and later elevated and exposed by erosion at the earth's surface.

Pothole. As used in this guide, a pot-shaped pit or hole eroded into bedrock by the erosive action of stones moved in a swirling fashion by fast-moving water.

Precambrian Eon. That period of earth history between its formation, some 4.5 billion years ago, and about 570 million years ago when life increased dramatically in both numbers and diversity. Strictly speaking my use of this term is incorrect. Precambrian time is divided into two eons, the Archean Eon, ending about 2,500 million years ago, and the Proterozoic Eon. For simplicity I have chosen to employ the term 'Precambrian Eon' and reduce 'Archean' and 'Proterozoic' to era status, neither of which I use in the text. My apologies to purists.

Precambrian shield. See **Canadian Shield**.

Pyroxenes. A group of closely related, dark-coloured, igneous silicate minerals composed of one or more of the elements calcium, sodium, magnesium, iron, chromium, manganese and aluminum.

Quartz. A mineral compound of silica and oxygen, most commonly colourless or white, originally formed during crystallization of molten magma and a common constituent of igneous, metamorphic and clastic sedimentary rocks.

Quartzite. A hard, unmetamorphosed sandstone, consisting chiefly of quartz grains cemented together by secondary crystalline quartz.

Quaternary Period. The Cenozoic Era is divided into two periods: the Tertiary (66 million years to 1.9 million years ago) and the Quaternary Period (1.9 million years ago to present). The Quaternary is further subdivided into the Pleistocene Epoch (1.9 million years to 10,000 years ago, and the Holocene or Recent Epoch which includes the present.

Recessional moraine. See **moraine**.

Recessive. See **Differential erosion**.

Resistant. See **Differential erosion**.

Rift. A fracture in the earth's crust caused by tensional forces such as those associated with the splitting of tectonic plates. A deep valley, such as the African Rift-Valley or the valley occupied by the Dead Sea, commonly forms along the fracture.

Ripple marks. Parallel, small-scale ridges and hollows produced by currents acting upon loose sediments such as those seen on modern beaches.

Rock flour. As used in this guide: fine powder formed by the abrasive action of stones, embedded in a glacier, upon the underlying bedrock.

Rocky Mountain Trench. Part of a system of linear, deep, fault-controlled valleys extending the entire length of the Canadian Cordillera.

Rodinia. The name given to a giant supercontinent incorporating all of the world's continents, formed about 1,150 million years ago and which broke up about 750 million years ago.

Sandstone. A clastic sedimentary rock composed of rounded to angular fragments of sand-size, set within a matrix of silt or clay, and cemented by calcite, silica or iron oxides. The sand particles most commonly consist of quartz, however, other minerals such as feldspars and micas commonly are included.

The distinction between sandstone and siltstone is one of grain, or clast size; clasts of siltstone have diameters less than 1/16 mm.

Seafloor spreading. The hypothesis which explains the creation of oceanic crust by convective upwelling of molten magma along the global mid-ocean **spreading ridge** system and the growth, or movement away from the ridge system of the newly forming crust.

Schist. See **metamorphic rocks**.

Shale. A fine-grained clastic sedimentary rock formed from the consolidation of clay and silt (lithified mud).

Slate. See **metamorphic rocks**.

Splay fault. A thrust fault that splays off from another thrust fault.

Spreading ridge. See **seafloor spreading**.

Strata. Plural of **stratum**. Layers of parallel sedimentary or volcanic rocks. **Stratified** refers to the condition of being layered in parallel **strata**.

Striations. As used in this guide: linear scratches or grooves carved into bedrock by stones embedded into the base of a moving glacier.

Strike. (1) The direction, or azimuth, of a horizontal line drawn on an inclined planar surface. (2) The direction of a linear feature such as the surface trace of a fault or a mountain range.

Strike-slip faults. Vertical fractures through the crust, the rocks on either side of which have moved horizontally past one another. Several such faults occur throughout the Canadian Cordillera, some of which separate blocks of crust that have moved several hundred kilometres with respect to one another.

Stromatolites. A structure produced by sediment trapping, binding or precipitation as a result of the growth of cyanobacteria. They have a wide variety of forms, from horizontal matts to columnar, domal or spherical shapes.

Stromatoporoids. Extinct marine, reef-building organisms of uncertain biological affinity, but thought by most paleontologists to be related to sponges. They secreted skeletons of calcium carbonate in a wide variety of shapes and were especially abundant from Ordovician through Devonian time.

Structural style. A general term describing the manner by which rocks of a given region have been deformed. For example, the "structural style" of the southern Rockies is one dominated by thrust faults, whereas the structural style of the Mackenzie Mountains is characterized mainly by folds.

Structural control. The influence of structural features such as **anticlines**, **synclines** and faults on the development of landforms.

Surface trace of a fault. The line of intersection of the plane of a fault with the earth's surface ("fault line").

Syncline. A down-fold in which the strata have been bent concave upward. The youngest strata are found in the centre of the fold.

Subduction. The process of consumption of oceanic crust along deep sea trenches whereby one piece of crust descends, or is subducted beneath another. An example occurs off our west coast where the oceanic rocks of the Juan de Fuca Plate, created at the Juan de Fuca Ridge, have spread away from the ridge through seafloor spreading, and are being consumed along the Cascadia subduction zone off the west coast of Vancouver Island, Washington, Oregon and northernmost California.

Subsequent stream. As used in this guide, a tributary stream formed after uplift of the mountains and which flows parallel to the trend of the ranges.

Superterranes. Assemblies of smaller terranes which amalgamated prior to their accretion to western North America. Western Canada is constructed from two such superterranes. The Intermontane Superterrane consists of several smaller terranes with names such as "Stikinia," "Quesnellia" and the "Cache Creek" and "Slide Mountain terranes" which amalgamated about 225 million years ago and which collided with North America about 170 million years ago. The Insular Superterrane, consisting of smaller crustal fragments called "Wrangellia" and the "Alexander Terrane", crashed into the previously accreted Intermontane Superterrane by about 100 million years ago.

Synclinorium. A large downward flexure with many smaller anticlines and synclines superimposed upon it.

Talus. Rock fragments of any size or shape eroded from and lying at the base of a cliff or steep rocky slope; talus slopes consist of angular rock debris formed by rockfalls and the continual spalling of material from higher cliffs.

Tarn. See **cirque**.

Tectonic. An adjective referring to the large-scale structural and deformational history of the earth's crust in a given region.

Tethys. The name given to a subtropical seaway extending from the globe-encircling Panthalassan Ocean across southern Asia and Europe during formation and initial break-up of Pangea.

Terminal moraine. See **moraine**.

Terranes. Parts of the earth's crust which preserve geological records different from those of neighbouring terranes. The boundaries between terranes are faults. Some appear to be comparatively thin sheets, whereas others are at least eighteen to twenty kilometres thick. Some terranes originated close to their present position whereas others were formed thousands of kilometres away.

Incipient thrust fault

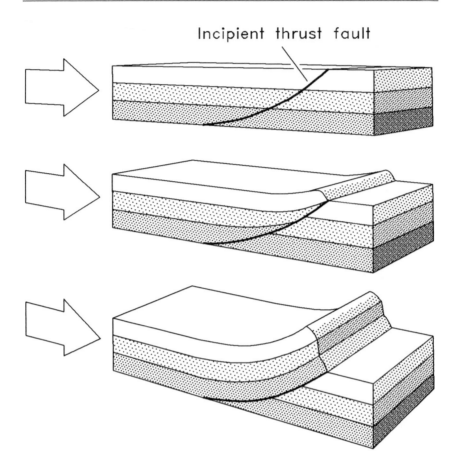

Thrust fault. A fault surface, generally inclined at less than 45 degrees, along which the rocks above the fault (hanging wall) have moved upwards and overtop of the strata below the fault (footwall). In the Canadian Rockies thrust faults separate thick sheets of strata (thrust sheets) which have been shoved eastwards many tens to hundreds of kilometres. In most cases in the Canadian Rockies the rocks immediately above the fault surface are older than those immediately below.

Till. Unsorted and unstratified glacial sediment deposited by and beneath a glacier, generally consisting of clay, silt, sand, gravel and boulders.

Trace faults. As used in this guide, a wide variety of markings left by bottom-dwelling, scavenging and burrowing organisms.

Trellis drainage. A drainage pattern characterized by parallel main streams intersected at, or nearly at right angles, by their tributaries, which, in turn, are fed by secondary tributaries flowing parallel to the main streams.

Trilobites. A segmented, three-lobed, bottom-dwelling, marine arthropod that lived during the Paleozoic Era but which was most abundant during the Cambrian and Ordovician periods. My favourite is *Ogygopsis klotsi*, the species named after a surveyor for the C.P.R., one Otto Klots.

Tufa. Spongy masses of porous, soft calcium carbonate, gypsum or silica commonly formed at hot springs. Dissolved carbon dioxide and hydrogen sulphide in hot-springs water escapes as the water cools, causing the precipitation of calcium compounds.

Turbidites. Sediments deposited from dense, sediment-laden currents. Turbidites commonly result from submarine landslides and are recognized by specific kinds of internal sedimentary structures such as graded beds and cross-laminations resulting from ripple-producing flow of the current.

Type-section. The locality where a formation's typical characteristics and lithology are best displayed and against which all other occurrences of the formation can be compared. Commonly it is the place from which the formation gets its name.

Unconformity. As used in this guide, a gap in the geological record (absence of strata) between two formations whereby the uppermost formation is substantially younger than the lowermost, thus indicating that an interval of earth history spanning the age limits between the two formations is locally unrepresented by rock strata.

U-shaped valleys. Glaciated valleys with a characteristic U shape formed by erosion and deepening of pre-existing mountain valleys.

Varves. Thin (1 mm to 1 cm), alternating light and dark layers of sediment that accumulate annually in glacial lakes. The light layers consist of sand and silt brought to the lake by glacial streams during the summer months. The dark layers are clay and organic material which remained suspended in the lake water until the surface froze, allowing these finer materials to slowly settle to the bottom during the winter.

Vugs. Large pores, or holes in carbonate rocks, commonly lined with crystalline minerals of different composition from that of the host rock.

Wisconsinan glacial stage. The last of several major advances of continental and Cordilleran ice sheets that covered much of the Northern Hemisphere between 80,000 and 10,000 years ago. Those affecting North America were called the Laurentide Ice Sheet and the **Cordilleran Ice Sheet**.

INDEX

(bold numbers indicate photograph or drawing)